MODERN ELECTRICAL EQUIPMENT
FOR AUTOMOBILES

MOTOR MANUALS VOLUME SIX

MODERN ELECTRICAL EQUIPMENT
FOR AUTOMOBILES

Arthur W. Judge

Associate of the Royal College of Science, London; Diplomate of the Imperial College of Science and Technology (Petrol Engine Research); Whitworth Scholar; Chartered Engineer; Member of the Institution of Mechanical Engineers; Member, Society of Automotive Engineers (U.S.A.), A.P.S.A. (U.S.A.)

SECOND EDITION

LONDON

CHAPMAN AND HALL

*First published 1962
by Chapman and Hall Ltd
11 New Fetter Lane, London EC4P 4EE
Second edition 1970
Reprinted 1973, 1975*

© 1962, 1970 *Arthur W. Judge*

*Printed Offset Litho in Great Britain by
Cox & Wyman Ltd, Fakenham, Norfolk*

ISBN 0 412 09700 1

*This limp bound edition is
sold subject to the condition that it
shall not, by way of trade or otherwise, be
lent, re-sold, hired out, or otherwise circulated
without the publisher's prior consent in any form of
binding or cover other than that in which it is
published and without a similar condition
including this condition being imposed
on the subsequent purchaser*

*All rights reserved. No part of
this book may be reprinted, or reproduced
or utilized in any form or by any electronic,
mechanical or other means, now known or hereafter
invented, including photocopying and recording,
or in any information storage or retrieval
system, without permission in writing
from the publisher*

PREFACE TO THE SECOND EDITION

The necessity for a reprint of the previous edition of this Manual has afforded an opportunity of bringing the information in certain parts of the book up to date, by the addition of a new Chapter 13 which deals with the more important developments that have occurred in the interim. This method has been adopted in order to simplify and to expedite the preparation of the present edition.

As with the other Manuals of the Series, the elementary method of treatment of the subject has been retained, but where considered necessary some theoretical aspects are discussed. The previous edition has been checked and where desirable certain minor alterations and improvements have been made in order to clarify the text.

There have been several important developments in electrical components and wiring methods since the last edition, the more interesting of which have included the wider use of electronics in the design and construction of certain automobile parts. Examples of these are the use of transistors, diodes and printed circuits on flat and flexible bases, notably for instrument panels, while miniaturized versions of printed circuits are finding wider applications in automobile components, e.g. for alternator voltage control units.

In order to assist the non-technical reader, for whom these Manuals were originally intended, a brief outline of the theory and applications of diodes and transistors has been included to help him to understand the circuits using these modern components. For those seeking further information some useful sources are given in the references at the end of Chapter 13.

Amongst the new material given in the new Chapter the following may here be mentioned: printed and miniaturized circuits; modern alternator design and performance; improved conventional and transistorized ignition systems; new sparking plugs and batteries; quartz iodine head, spot and fog lamps; improved flashing indicator systems; multi-speed windscreen wipers; magnetic-pulse type speedometers and tachometers. An account is given of a new type of flexible optical cable which transmits light from a single source at one end to illuminate several distant small areas, e.g. on the instrument panel and on switches. In conclusion one would like to take this opportunity to acknowledge the co-operation in the preparation of this revised edition of the following individuals and manufacturers, namely H. H. Jones, for his advice and useful suggestions, T. C. Hoskins (Joseph Lucas Ltd.), B. Summerfield (AC–Delco Division of General Motors Ltd. England) and D. A. Cooke (Smiths Industries Ltd. Motor Accessory Division).

ARTHUR W. JUDGE

Farnham, Surrey
1969

CONTENTS

1. AUTOMOBILE ELECTRICAL SYSTEMS — *page* 13
2. THE STARTING SYSTEM — 31
3. THE CHARGING SYSTEM — 67
4. IGNITION SYSTEM PRINCIPLES — 125
5. COIL AND OTHER IGNITION SYSTEMS — 152
6. THE MAGNETO — 195
7. THE SPARKING PLUG — 211
8. THE AUTOMOBILE BATTERY — 233
9. THE LIGHTING SYSTEM — 261
10. AUTOMOBILE ELECTRICAL INSTRUMENTS — 292
11. MISCELLANEOUS ELECTRICAL EQUIPMENT — 316
12. WIRING AND INSTALLATION — 346
13. LATER DEVELOPMENTS IN ELECTRICAL EQUIPMENT — 359

INDEX — 406

ENGLISH – AMERICAN GLOSSARY

AUTOMOBILE TERMS

English	*American*
Aluminium	Aluminum
Anti-clockwise	Counterclockwise
Bush (metal)	Bushing
Carburettor *or* Carburetter	Carburetor
Car Bonnet	Hood
Colour	Color
Contact Breaker	Breaker
Current Regulator	Current Regulator *or* Limiter Relay
Cut-out	Cut-out Relay
Dismantling	Disassembling
Dynamo	Generator
Earth (metal)	Ground
Fibre	Fiber
Gauge	Gauge *or* Gage
Headlamp Aligners	Headlamp Aimers
Inlet Manifold	Intake Manifold
Instrument Panel	Instrument Cluster Assembly
Licence Plate	License Plate
Lb. per sq. in.	PSI
Motor Car	Automobile
Moulded	Molded
M.P.H.	MPH
Paraffin	Kerosene
Petrol	Gasoline
Reversing Light	Back-up Light
R.P.M.	RPM
Screwed	Threaded
Spanner	Wrench
Sparking Plug	Spark Plug
Starting Motor	Starter *or* Starting Motor
Vice	Vise
Voltage Regulator	Voltage Regulator *or* Limiter Relay
Windscreen	Windshield
Wing	Fender

CHAPTER I

AUTOMOBILE ELECTRICAL SYSTEMS

THE modern electrical system has been developed, over a period of some fifty years from the days of the early motor-car which usually had only one electrical system, namely, that of the ignition comprising either a trembler coil and battery or a magneto. The replacement of the magneto by the coil ignition system with its necessary battery unit, necessitated some means of keeping the battery charged and this brought the dynamo into more general use. Having a regularly charged battery the earlier advantage taken of this unit was to provide electric current for the headlamps and tail lamps and, later, to the electric motor starting unit that relieved the starting handle of most of its duties.

From these early beginnings, the modern more complex automobile electrical system has been developed. This system involves the use not only of purely electrical devices, but also mechanical and optical ones, so that the modern motor engineer and mechanic must be familiar with the principles of electrical, mechanical and optical subjects.

Basic Principles

The present-day automobile electrical system may be regarded, broadly, as a transportable power station consisting of a petrol or Diesel engine, a part of the output from which is employed to drive an electric generator—which will be here termed the dynamo—to provide the source of electricity. Since it is necessary to draw electric power from the system at low road speeds and also when the car is at rest with its engine stopped, there must evidently be some kind of electrical energy storage unit, namely the accumulator or battery. Thus, the engine-driven dynamo keeps the battery charged under all road running conditions, and the various electrical units of the automobile take their current supply from the battery.

The electrical system may be likened to that of an hydraulic system, consisting of a power-driven pump which draws water from a lower storage tank and pumps this water up to a reservoir or accumulator at a much higher level, thus creating a 'pressure head' which provides hydraulic power for operating various purposes, e.g. hydraulic motors for power purposes, hydraulic circuits, cooling systems, etc. It will thus be apparent that the reservoir acts as a power storage device in the same manner as an electrical accumulator and must be maintained at its required pressure head and water quantity by a power-driven pump. In this hydraulic parallel the pressure head and water quantity available are analogous to the electrical pressure, or voltage and the current supply.

Automobile Electrical Systems

To the beginner, the electrical wiring diagram of a modern automobile is apt to prove rather confusing, with its many electrical components and numerous circuit-indicating lines. This is due largely to the relatively large number of electrical components with their cables and also to the fact that invariably all of these components have a common origin of electrical supply, namely, the battery. As it would, of course, be very inconvenient to connect every electrical item to the battery terminals, instead, connections are made to certain other components, such as the voltage or voltage and current regulator of the dynamo, or to a distributor unit conveniently mounted for making the necessary connections. The battery is therefore provided, as a rule, with two main cables leading to the heaviest current consumer, namely, the electric motor used for rotating the engine crankshaft, *via* the flywheel, for starting purposes. The live lead or cable from the battery is taken to the starting motor switch, from which it goes, to the ammeter terminal, from the other terminal of which a lead is taken to the ignition switch and thence to the distributing device, i.e. the dynamo regulator unit and/or, a separate distributing component which supplies live current, through suitable switching units to the various electrical circuits that make up the complete electrical system.

There is one important point that should always be remembered when studying automobile circuits, namely, that *all electrical components* which draw their current from the battery *are always connected in shunt (or parallel) with the battery via the live lead to the current distribution unit*. Further, with the possible exception of the high tension (voltage) ignition system and the car radio—both of which systems are initiated by battery voltage, all the electrical components of automobiles are operated at battery voltages, i.e. 6, 12 or 24 volts.

Insulated and Earthed Return Systems

In the case of domestic and industrial wiring systems all components are connected or wired with twin insulated cables, the mains voltages being of 230–240 volts. In the earlier automobile wiring systems fully insulated cables were used for the electrical components, but latterly this method has been wholly replaced, for automobiles, by the single wire, or *earth return* method. Thus, a single insulated, or live cable conveys current from the insulated terminal of the battery, while the metal members of the engine and chassis provide a common return path for the various components, to the earthed terminal of the battery.

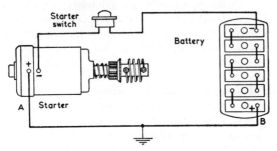

FIG. 1. Schematic arrangement of car starting system

Referring to Fig. 1, which shows, schematically, the simple circuit of a car engine starting motor, the connecting lines denote insulated cables. If, however, the return current conductor AB is connected to the metal of the car, this will not affect the flow of the current, so that the return cable can be dispensed with if the positive terminals of the starting motor and battery are connected to their nearest metal or earth parts on the chassis. Thus, by using the common metal return, one set of insulated cables can be dispensed with. The advantages of this 'earth return' system are that (1) The resistance of the earth return part of the circuit—provided sound metal connections are made, throughout—is almost negligible, whereas that of the insulated return cable is a definite quantity. (2) For this reason, it follows that the cross-section of the copper conductor of the earth return cable need only be one-half that of the insulated return system, so that for the single earth return system the weight of cable is only one-half that of the insulated or twin cable. Therefore, the earth return system is more economic. (3) The earth return system, since it dispenses with one cable, gives a simpler and less bulky wiring system, in practice.

The principal disadvantage of the earth return system is that of possible faults or breakdowns in circuits owing to bad metallic connections, due to loose screws or corrosion. In this connection the insulated system is generally accepted as being more reliable under road shock and vibration conditions and it is for this reason that it is still used on many commercial and passenger vehicles. Since the electrical equipments are relatively simple in these vehicles the extra cost factor for the cables is not so important.

In the insulated system a single failure to the metal earth of the vehicle does not necessarily put the electrical system out of action, but only the component concerned, whereas in the earth return system if the insulated cable becomes earthed the whole system is apt to fail—if only because the battery discharges.

It is more difficult to trace the origin of wiring faults in the insulated than in the earth return system, so from the practical point of view this is another advantage for the latter method of wiring.

Understanding Electrical Systems

While some information on this subject has already been given, it was necessary to digress for a while in order to explain the earth return wiring system now standard practice on automobiles and certain other vehicles, not in the motor-car category. The use of this system certainly simplifies the electrical wiring diagrams,

AUTOMOBILE ELECTRICAL SYSTEMS

although these can still be complex, except in the case of miniature cars and small vans.

The complete electrical system may be more readily understood when it is realized that it consists of *five basic systems*, as follows: (1) *The Engine Starting Motor System*. (2) *The Ignition System*, (3) *The Battery Charging System*, (4) *The Lighting System*, and (5) *The Miscellaneous Electrical Equipment System*.

Each system takes its current supply nominally at constant voltage either directly or indirectly from the battery. Sometimes as in the case of the starting, horn or direction indicator circuits the supply is required intermittently, whilst for the ignition system the electrical demand is continuous, so long as the engine is operating—in the case of petrol, but not Diesel engines.

It is proposed, now, to consider briefly each of the individual circuits commencing with the engine starting motor system.

(1) The Engine Starting Motor System. In modern cars the starting-handle method has been replaced by that of the starting motor. This method obviates the use of physical effort on the part of the driver and does not require him to get out of the vehicle.

The starting motor, which is of sufficient electrical output to crank the engine at a satisfactory speed to enable it to start under cold weather conditions, is of the high starting torque kind—usually of the four-pole series type—and is wired in series with the operator's switching system and the battery, as shown, schematically, in Fig. 2, from which it is seen that when the switch S is closed current from the battery is directed to the starting motor $S.M.$ which is provided with a small pinion gear

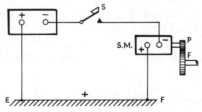

FIG. 2. Simple circuit diagram for starting motor

P meshing with a larger annular gear, integral with the engine's flywheel *F*. The battery current flows through the starting motor to the metal frame of the chassis; since one pole of the battery is also connected to this frame—or the *common earth*, or *ground*—the electrical circuit is completed and the motor rotates the engine's crankshaft until the engine starts, when an automatic device in the starter pinion disconnects the pinion from the flywheel gear ring.

Since the engine operates at a much lower speed than the starting motor and as it is necessary to provide the maximum starting torque, the use of the relatively large flywheel gear and small motor pinion give the desired speed reduction and torque. Usually the reduction gear ratio is from 8 : 1 to 12 : 1, although higher ratios, up to 16 : 1 are occasionally used; in this case the overall dimensions of the motor can be made smaller.

It will be observed, from Fig. 2, that the circuit contains no ammeter, the reason being that it would require an ammeter having a scale calibrated to show several hundred amperes —which is the usual initial starting current value—whereas all of the other electrical components, and circuits require very much smaller currents. Actually, the maximum current, excluding the starting circuit is that of the dynamo charging supply, namely, from 20 to 30 amperes in most British cars, but rather higher values in American cars. Thus, an ammeter reading to a maximum of about 30 amperes to 50 amperes, according to the type of car, will cover requirements.

In regard to the switching system, due to the heavy currents concerned the simple type shown, schematically, in Fig. 2, cannot be used. Instead an ordinary switch, mounted on the dashboard, and operated by the driver is used to actuate a relay which brings into operation a heavy amperage type of switch, that is generally mounted on the starting motor casing.

Since the subject of starting motors and their circuits is dealt with in Chapter 2 it is unnecessary to dwell further on the starting system, here.

(2) The Ignition System. This system is necessary with petrol engines in order to provide the high voltage sparks, at accurately timed intervals, for the purpose of igniting the petrol-air mixture within the engine's cylinders.

Diesel engines do not require electric ignition systems since the temperature of the air at the end of the compression stroke is sufficient to burn the fuel which is then injected. However, in certain types of Diesel engines, known as pre-combustion chamber engines, it is necessary to employ special plugs, known as *heater* or *glow plugs* to heat the air sufficiently for starting the engine from cold.

The other method of igniting the petrol-air mixture that does not require current from the battery and therefore needs no special low voltage circuit is that of the *magneto*, which generates its own low voltage current and from this produces the high voltage sparks required for ignition. This self-contained ignition unit requires an engine drive and high tension cables to connect its high voltage contacts to the sparking plugs.

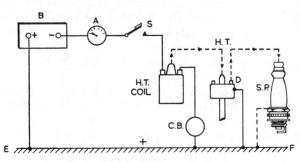

FIG. 3. The ignition system circuit

As mentioned earlier, the magneto has been replaced on automobiles and most other vehicles, except certain petrol-engine tractors, by the coil-ignition system shown, schematically, in Fig. 3, in which the battery B supplies low voltage (tension) current, *via* the ammeter A and ignition switch S to the primary, or low tension circuit of a special transformer coil, known as the high tension or H.T. coil. The purpose of this coil is to convert the low tension current, from the battery at 6 or 12 volts, to the high voltage—usually about 8,000 to 12,000 volts, required to produce the high tension sparks within the engine cylinder, for mixture ignition purposes.

The current from the low tension circuit of the H.T. coil flows from the outlet terminal to a device known as a contact breaker, shown at *C.B.* in Fig. 3, and thence to the common earth *EF* and back to the battery. The purpose of this contact breaker—as is explained in detail in Chapter 4—is to interrupt, or break the low tension circuit every time a spark is required at the sparking plug.

The high tension circuit from the H.T. coil to the sparking plug is indicated by the dotted lines, from which it will be seen that the high tension current goes from the central terminal of the H.T. coil to the upper part or insulated cap of the contact breaker unit *D* known as the *distributor*. This consists of a rotating metal arm which directs the high tension current to each of the sparking plugs *S.P.*, in turn, in the case of a multi-cylinder engine. As with the low tension current from the contact breaker, the high tension current after bridging the sparking plug gap flows to the outer metal shell of the sparking plug—the central conductor of which is insulated from this shell—and thence to the metal frame of the engine and thus to the common earth *EF*.

It should be mentioned that in our present considerations we are concerned only with the low tension circuit, from the viewpoint of the basic wiring or circuit diagrams, but it may be of interest to study the complete circuit diagram shown in Fig. 86 in Chapter 5 for a typical four-cylinder engine ignition system.

H.T. Coil Consumption. It may be of interest to note that the usual value of the current, from a 12-volt battery, that flows through the low tension coil of the ignition circuit usually lies between 3·5 and 4·5 amperes when the contacts are closed and the engine is at rest, falling to about one-half of these values when the engine is operating.

(3) The Battery Charging System. Since the whole of the electrical supply for the various automobile components is derived from the battery it is important to maintain the battery in the charged condition all the time the car is in use. This is done with the aid of the dynamo and certain devices to protect the circuit against excessive voltages and the dynamo coils against battery current discharge when the engine is idling or at rest.

There are occasions, however, when current is required from the battery while the engine is switched off, as when a car is parked at night with its side and rear (parking) lights switched on. In such cases, the battery capacity is sufficient to cope with the slow rate of discharge over a period of many hours, assuming the battery is in good condition. Basically, the battery charging circuit consists of a number of components wired as shown schematically in Fig. 4. The battery B is supplied with current

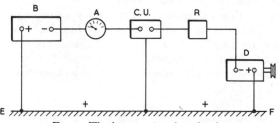

FIG. 4. The battery charging circuit

from the dynamo D, which is driven by belt and pulley from the crankshaft of the engine, usually at 1·5 to 2·5 times engine speed. From one dynamo terminal the current is taken to a regulator unit R, the purpose of which is to maintain the voltage of the output constant within certain prescribed limits. Thus, with a typical British car dynamo operating at 2,000 r.p.m. the voltage is 13·5 and the current 20 amps. The battery voltage being nominally 12·0, it will be understood that the dynamo will maintain a good charging current when required. The regulator, however, provides for varying the current value according to the state of the battery. Thus, when the battery charge is 'low', the regulator arranges for a higher current value and when fully charged, a low value—even although the dynamo may be operating at a high speed.

The other electrical component marked *C.U.*, in Fig. 4, is known as the *cut-out* and its purpose is to disconnect the battery from the dynamo-regulator system when the dynamo is running at too low a speed (or voltage) to supply any charging current and to re-connect the battery when the dynamo speed has increased sufficiently to supply current at a higher voltage. Thus, in a typical instance when the engine is at rest or idling, the dynamo will not supply sufficient energy to charge the battery and so the cut-out remains in the circuit-breaking position. When, however, the dynamo is accelerated to a speed of 1,050 to 1,200 r.p.m. at 13·0 volts the cut-out operates to close contact points, so that the dynamo, with its regulator is connected to the battery, *via* the ammeter. In all modern cars the cut-out is mounted on the same unit as the regulator coils; the latter are designed to control the voltage only, or both the current and voltage as in American and, more recently, European practice.

It will be observed, from Fig. 4, that the dynamo, regulator and cut-out members have earth returns as indicated by the *EF* common return connections in this diagram. The charging or discharge current is indicated by the ammeter *A*. Internal wiring diagrams and other particulars of the battery charging system are given in Chapter 3.

(4) The Lighting System. This system comprises all of the lamps used for illumination, pilot lights, indicators and for warning purposes, e.g. the rear red brake lights.

The principal lamps include the headlamps, side and tail (parking) lamps, the flashing-type direction indicator lamps, car interior panel and illumination lamps, the ignition warning red lamp, oil pressure indication lamp, headlamp beam (upward) indication lamp, etc.

The complete lighting circuit is made up of a number of individual circuits for the single lamp or pairs of lamps, each with its own switch, 'live' connection and earth return. It is the individual wiring of the various lamps that makes the complete wiring diagram appear to be rather complicated, but if the basic wiring principle of the individual lamps is remembered it much simplifies the tracing of the different lamp circuits.

Fig. 5 shows a schematic layout of both the lighting and accessories circuits; the accessories are here assumed to include the electric horn *H*, windscreen wiper *W*, car heater *C*, instruments, clock, car radio *R*, etc. Some of these are included in the diagram; they are all connected in parallel or shunt across the battery terminals or to the 'live' terminals of other components which are connected by cables, *via* the ammeter, to the battery 'live' terminal.

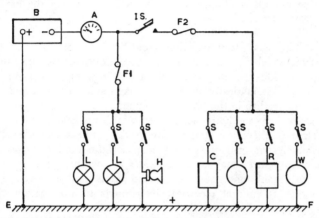

FIG. 5. Schematic layout of the lighting and accessories systems. *Note.* The letters L, H, C, V, R and W, denote the lamps, horn, car heater, voltmeter, radio and windscreen wiper

In Fig. 5 the battery *B* supplies current through the ammeter to the lighting circuits as indicated by the lamps *L* with their individual switches *S*. Usually, these circuits, which are always available for operation whether the ignition switch *I.S.* is 'On' or 'Off', are protected by a single fuse *F*1, which in a typical car would be of the 40 to 50 amp. kind.

The various electrical accessories protected by the other fuse *F*2—which is of the 25 to 35 amp. kind—can only be operated when the ignition switch *I.S.* is 'On'. Some of these accessories, such as the instruments, e.g. pressure, temperature and fuel level gauges, have no separate switches, so that they are always in operation; similarly, the ignition circuit is always 'live' when the ignition switch is 'On'. Other components, such as the electric horn, windscreen wiper, car heater and radio, have individual operating switches, since these accessories are only required to operate intermittently.

In regard to the lamp switches it is usual to provide a single switch unit for the headlamps and side and tail (parking) lamps, such that the first 'On' position of this switch puts the side and tail lamps 'On', while a further movement of the switch causes the headlamps to illuminate. This method obviates the use of separate switches.

(5) The Miscellaneous Electrical Equipment System. This system, which includes all the electrical components other than those previously dealt with under the headings (1), (2), (3) and (4), is shown schematically on the right hand side of Fig. 5 and has been referred to in the previous section.

Apart from the horn, windscreen wiper, car heater, radio and electric petrol pump (when fitted) this system includes also certain electric-principle gauges, such as those fitted for recording the oil pressure, engine cooling water temperature, fuel level in the fuel tank and, sometimes, the crankcase oil level. The ammeter is another electrical instrument coming within this category.

The Complete Electrical System. The complete schematic wiring diagram, representing the combination of the five basic circuits, described earlier, is shown in Fig. 6, from which the individual circuits can readily be traced. In this diagram the common earth return circuit is indicated by the hatched line *EF* while the insulated 'live' cables are shown by the finer lines marked 'a'. It should, however, be pointed out that since none of the internal wirings of the components, e.g. the dynamo, motor and ignition coil are shown and, further, only a few of the relatively numerous electrical accessories, on the right hand side of the diagram are indicated, the actual wiring diagram is more complex than the schematic diagram, as will be evident from the example shown in Fig. 7 which is based upon the electrical system employed upon one of the Ford Prefect four cylinder engine cars, but modified to show the flashing light indicators.

This diagram shows in a clear manner the various major

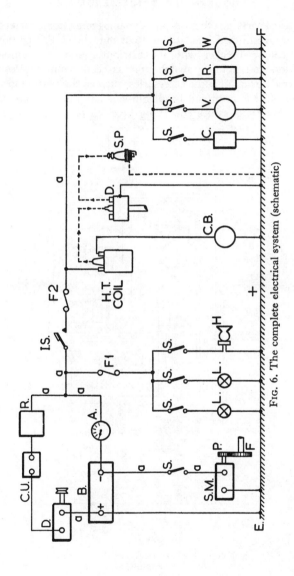

FIG. 6. The complete electrical system (schematic)

electrical parts, e.g. the two-pole dynamo (generator), starter motor, battery, combined regulator and cut-out, the H.T. (ignition) coil and contact-breaker with its distributor and the combination ignition and lighting switch unit as seen from the rear. The internal wiring of some accessories, such as the petrol tank gauge units, electric horn and headlamp-side lamp units is also shown.

The various cables conveying 'live' currents are, in practice,

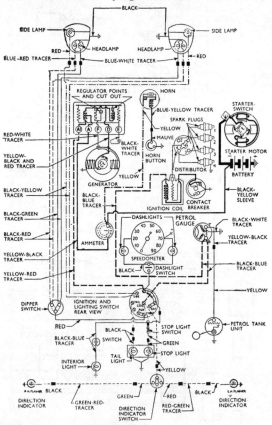

FIG. 7. Typical complete electrical circuit diagram for four-cylinder engine car

AUTOMOBILE ELECTRICAL SYSTEMS 27

provided with coloured braidings in order to distinguish them on the car; the colours and colour-combinations used in the present example are indicated alongside the cables.

In connection with the diagrams shown in Figs. 6 and 7, the method of showing which parts or terminals of electrical components are connected to earth (or ground), i.e. to the metal of the chassis frame or engine, is denoted by EF in Fig. 6 and by similar symbolic earth lines near to the various components in Fig. 7.

Fuller information in regard to the various components shown in Fig. 7 are given in the appropriate Chapters of this book, while more detailed particulars of automobile wiring systems are given in Chapter 12.

Summary of Electrical Applications. It is now possible to summarize the information given previously, under the five separate headings, in a convenient and easily understood manner, by the diagram shown in Fig. 8, which shows how the mechanical power supplied by the engine is utilized to supply all of the electrical energy required by the electrical components of the automobile. This diagram serves also to show the different energy conversions that occur in the electrical system.

The mechanical energy of the petrol or Diesel engine is derived from the combustion of the fuel-air mixture, this being in the nature of a chemical process. The engine supplies mechanical energy to operate the dynamo, which in its turn provides electrical energy to charge the battery. The charging process represents a conversion of electrical to chemical energy which is stored within the battery and taken from the battery when required, in the form of electrical energy to operate the electrical components. In some of these components, e.g. the lamps, heating system and sparking plugs the electrical energy is converted into heat energy while in others, e.g. the starting motor, horn, windscreen wiper, electric petrol pump and other power-operated units, the conversion is from electrical to mechanical energy.

Reverting to Fig. 8, the dynamo supplies charging current to the battery which, in turn provides the electrical supply for the five basic electrical systems described previously. The supply to

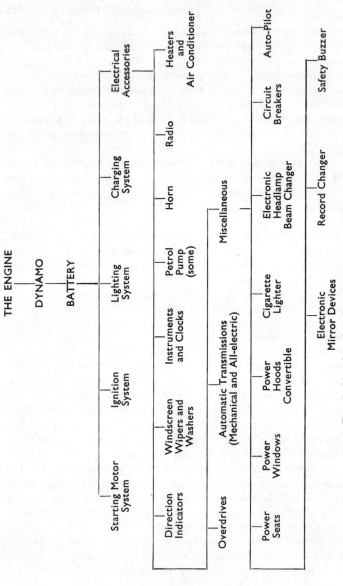

FIG. 8. Showing electrical energy analysis for automobiles

the 'Accessories' section is sub-divided into the common electrical components of most standard British cars, as denoted by the second row of sub-headings, commencing with 'Direction Indicators'. Included in these items is the *Overdrive*, or twin top gear of certain cars fitted with three- or four-speed gearboxes. This extra gear ratio is brought into operation, automatically as a rule, by an electrical circuit which when switched 'on' causes a solenoid to become energized, thereby moving a core member to actuate the gears.

Among the other electrical components used on most American and a few British luxury cars, are those shown in the bottom row and to some of these further details are given, later, in this volume.

Mention should here be made of an entirely electrical type of automatic transmission, namely, the Smith Autoselectric, which utilizes magnetic powder couplings with electric coil systems to actuate the mechanical gear systems. This transmission is supplied with its electrical energy by the battery.

Negative and Positive Earthing. In connection with the earth return method, wherewith only one insulated cable is required, it was hitherto the practice in this country to connect the negative pole of the battery to earth, so that positive current was supplied to the electrical components. This method is still the standard practice in the U.S.A.

The negative earth system was replaced in this country, about 1938, by the positive earth method, which on examination was shown to possess some definite advantages over the previously used system. These advantages largely concern the corrosion of certain parts and temperature of the sparking plug electrode. In connection with the corrosion factor it is well known that the positive or anode pole of a lead-acid battery is attacked by the liberated gases and if this is the 'live' pole, i.e. the negative pole earthed, and moisture is present the exposed part of the anode will become corroded; usually the positive pole with its cable connector was well greased to protect it against such corrosion.

It is also known that in the case of a pair of sparking points, separated by a suitable gap, the positive point corrodes or wears away more quickly than the negative point. Similarly, in the case of the sparking plug, if the central electrode is positive and the metal shell of the plug, negative, the electrode will wear away at a greater rate than the metal electrode of the shell gap. However, if the metal shell is made positive, by earthing the positive pole of the battery, the central electrode will have a much longer life.

There is another factor in regard to the sparking plug, namely its temperature. Thus, the voltage required to bridge the plug's sparking gap depends upon the temperature of the negative electrode and the hotter this electrode is, the lower will be the voltage. The effect of making the central electrode negative is to reduce the voltage needed to produce the spark, for this electrode is hotter than the metal shell electrode. Further, it has been shown that more uniform voltages at the sparking gap are obtained with the central electrode negative.

In the case of the metal rotor arm of the H.T. current distributor, if made negative to the fixed contacts in the distributor cap—which convey the H.T. current to each sparking plug in turn—it will wear away at a slower rate than if it were made positive.

An additional advantage of the positive earth method is that in the H.T. coil elements the primary circuit voltage is added to the secondary or high voltage circuit voltage, which is a more economical method, than if the coils are in opposition.

More Recent Negative Earth Systems. With the more recent adoption of the alternating current generator, with its advantages over the orthodox direct type, it has been found advantageous to employ the negative earth method, in connection with the A.C. current rectifier, incorporating transistors and diodes, so that there has become a reversion to the negative earth method. The previously mentioned important advantages of the positive earth for the ignition system are, however, retained by making the 'earth' of the system, only, positive. To clarify the matter of the ignition connections, the ignition coil and contact-breaker terminals are marked with their correct positive and negative signs, so as to ensure that the earth or return circuit is positive and the 'live' connection from the switch, negative.

CHAPTER 2

THE STARTING SYSTEM

IN comparison with the other electrical systems of an automobile, that of the starting motor and its switch and wiring is the simplest consisting, as it does, of switches, starting motor and battery, wired in series. Before dealing with the starting system and its electrical components it is necessary to consider the basic requirements of the starting motor from the point of view of the engine operation, under starting conditions.

Starting Torque and Power. The amount of power required to start an engine will depend upon its initial resisting torque and, once the crankshaft begins to turn, upon the speed at which the engine can be 'cranked'. It should here be pointed out that once the crankshaft is rotating its resistance to turning is much lower.

In the days of the starting handle method of starting an engine, it was necessary to exert a certain force on the end of the handle, in order to obtain the required starting torque and then to rotate the crankshaft at a certain minimum speed in order to start the engine. Thus, in the case of a small engine a force of about 23 lb. at the end of the 9-in. handle would produce a torque of $23 \times 9/12$, or $17\frac{1}{4}$ lb. ft. Further, if the crankshaft could be turned at 60 r.p.m. to start the engine the horse-power required can be calculated from the following relationship:

$$\text{Horse-power} = \frac{2\pi \times \text{torque} \times \text{r.p.m.}}{33,000} \text{ where } \pi = 3 \cdot 1416$$

In the present example it can be shown that about 0·2 h.p. would be exerted in starting the engine. Therefore, if an electric motor, with suitable gearing were used to start the same engine it would have to exert rather more than this value in order to overcome, also, the mechanical losses in the gearing. Assuming that the simple spur gearing has a mechanical efficiency of 90 per cent, then the starting motor would have to produce $\dfrac{0.2 \times 100}{90}$, or about 0·22 h.p.

Usually, the motor exerts its maximum output at a speed of 1,000 to 1,800 r.p.m. but, assuming an average speed of 1,100 r.p.m. the reduction ratio of the gearing between the motor armature shaft and the engine flywheel ring gear would be $\dfrac{1,100}{60}$, or 18 : 1, approximately. In practice starting motors develop their greatest horse-power at higher speeds for the small output motors and lower speeds for high output motors, such as would be used for starting automobile Diesel engines.

Engine and Starting Motor Torque Terms. The following are the usual terms employed in automobile engineering:

(a) *Engine Breakaway Torque.* This is the torque required to start moving the crankshaft from rest.
(b) *Engine Resisting Torque.* Once the engine crankshaft has started to move the torque required to keep it moving is known as the resisting torque. It is appreciably less—usually about one-half—than the breakaway torque.
(c) *Motor Locked Torque.* This is the torque developed immediately the battery current is switched on, so that the armature starts to rotate from the 'at rest' position.
(d) *Motor Driving Torque.* This is the name given to the motor torque when the armature shaft pinion gear is driving the flywheel through its meshing ring gear.

It is necessary for the locked torque to exceed the breakaway torque for the engine crankshaft to commence rotating and for the driving torque to exceed the resisting torque, for the motor to accelerate the engine, but as the engine speed increases, so does the resisting torque. Eventually the driving and resisting torques become equal and then the engine cranking speed attains its **maximum value.**

THE STARTING SYSTEM

Factors Affecting the Starting of Engines. The starting torque required for any petrol engine is influenced by a number of factors which may be considered, briefly, as follows:

Temperature. It is well known that a warm engine can be started by hand more easily than a cold engine, assuming ignition and carburation conditions are satisfactory. When the outside temperature is low, e.g. 0° C (32° F), not only is the viscosity of the engine lubricating oil much greater, so that the resisting torque is increased, but the carburation conditions for starting become more difficult. Thus, owing to the low cranking speed the carburettor supplies much more air than when operating under normal speed conditions. It is necessary, therefore, to enrich the mixture by reducing or 'choking' the air supply; some of the fuel drawn into the cylinder becomes deposited on the cold cylinder walls and helps to lower the viscosity of the oil film. Thus, after a short period of cranking the engine tends to rotate at a rather higher speed.

Lubricating Oil. Previously, most engine oils in order to attain the desired low viscosity condition under normal hot engine speeds, were rather too viscous when cold, thus increasing the resisting torque.

Fig. 9 illustrates some experimental torque values for various grades of oil for a less recent design of automobile engine, at 2° F (30° F). below freezing point The dotted curve E indicates the starting motor's torque-speed graph, with a starter pinion gear to flywheel gear ratio of 1 : 10. In the case of the four curves A denotes a competition engine oil, B a summer grade oil, C a winter (or less viscous) grade oil and D a grade of oil used under winter conditions on American cars of the same period.

The cranking speed of the motor is that at which the starting motor torque and the resisting torques are equal, and is therefore shown on the graphs by the intersections of the motor and engine resisting torques. Thus, for the grades of oil represented by the graphs A, B, C and D the small circles indicate the cranking speeds.

The results indicate that the cranking speeds increase with a

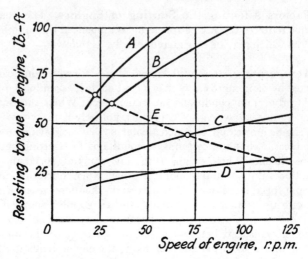

FIG. 9. Petrol engine resisting torques with lubricating oils at 30° F

decrease of oil viscosity. Thus for the winter grade oil curve C, the cranking speed is 73 r.p.m. while for the summer grade it is only 30 r.p.m.

It has been found possible to establish an approximate relation between the resisting torque of a petrol engine and the engine speed and oil viscosity. Thus:

$$\text{Resisting torque} = k \sqrt{\text{speed} \times \text{viscosity}}$$

Ignition Requirements. Other factors remaining satisfactory for engine cold starting, it is important that:

(1) The ignition system generates a sufficiently intense spark to ignite the compressed air and petrol mixture under low temperature starting conditions. In this connection the camshaft-type magneto is superior to the ordinary magneto, while a properly designed coil ignition system is superior to either of the magnetos, since it gives a higher voltage spark at speeds just above zero, i.e. from 30 to 70 r.p.m., engine speeds.

(2) The spark must occur at the correct moment, before the end of the compression stroke. For slow running the spark should

occur earlier than for normal running conditions. Thus, in earlier ignition systems a driver's control was provided to retard the ignition for starting and advance it, progressively, for speed increase. In modern ignition systems having centrifugal governor control of the ignition timing the ignition point is retarded, automatically, to its correct position for cold starting.

Carburation. Mention has been made of the necessity to choke the (main) air supply to the carburettor for cold starting purposes. This is usually done with the aid of a kind of butterfly valve located in the main inlet air passage and operated by a one- or two-position control on the instrument board. Modern carburettors are designed so as to have good starting and idling characteristics, but excessive use of the choke, should the engine not start for any reason, other than carburation, tends to fill the cylinders with an over-rich mixture which may make subsequent starting difficult. A common cause of difficult starting due to this cause is that of deposition of petrol between the insulated and earthed electrodes of the sparking plugs; this is more especially the case with two-cycle engines. The remedy is to remove and dry the plug points and internal insulation. The engine should be cranked with the throttle full open and ignition switched off, to get rid of the over-rich mixture, before repeating the starting procedure. Automatic chokes largely overcome hand choke troubles.

Some Other Factors. Apart from the previous factors, the engine starting torque required will depend upon the design of the engine, condition of the piston and its rings, the cylinder bore, the state of the connecting rod and crankshaft bearings, since these items determine the internal friction which has to be overcome. Thus, a new engine with its smaller bearing clearances and tighter piston assembly will require a greater starting torque than a worn engine with its freer bearing surfaces.

Another factor with automobile engines is the compression pressure which, in modern engines is relatively much higher than for previous engines and therefore needs a somewhat greater effort to crank the engine over its successive compression strokes.

Again, if the lubricant in the gearbox is fairly viscous at low temperatures, the gearbox main shaft gears will contribute to the starting and resistance torques. In such cases, a temporary expedient is to try declutching the gearbox shaft during the starting motor operation.

Diesel Engines. In general the automobile Diesel engine is appreciably more difficult to crank than the petrol engine, owing to its much higher compression pressures. Thus, the usual compression ratios range between about 14 : 1 and 18 : 1, compared to the ratios of about 7·0 : 1 to 10 : 1 for automobile petrol engines. The cylinder-piston friction is relatively greater, owing to the use of more piston rings and to the greater thrust forces.

From the combustion point of view, it is necessary to attain a certain high temperature in compressing the air charge, before the injected fuel will ignite. When the engine is cold it is obviously more difficult to attain this temperature, namely, about 370° C to 450° C, so that pre-heating of the air by various methods is employed in many instances.

It can be stated that automobile *Diesel engines of a given cylinder capacity* have relatively higher starting and resisting torques and, therefore, require more powerful starting motors than for petrol engines. Further, from the viewpoint of ignition temperatures, Diesel engines must be cranked at higher speeds. Usually, the cranking speeds are of the order, 100 to 200 r.p.m. at low temperatures. It is often necessary to crank these engines for longer periods, namely, 30 sec. to 1 min. or so, if no auxiliary starting devices are fitted.

The necessity to employ larger output starting motors also entails the use of larger capacity batteries. Fuller information on this subject will be found in the footnote reference.[*]

Starting Motors. The three principal kinds of electric motor, using direct current, in use, today, are (1) the Shunt-wound, (2) the Series-wound and (3) the Compound or Series-shunt wound. Of these, both (1) and (3) are employed for automobile engine cranking purposes. The second or shunt-wound motor is practically a constant speed machine, there being only a few per cent variation in speed between no-load and full-load.

[*] *High Speed Diesel Engines.* A. W. Judge (Chapman & Hall Ltd.).

(1) The Series-wound Motor. In this type (Figs. 10(2) and 16) the field windings around the pole pieces are connected in series with each other and with the armature, so that the same current is carried by the windings and armature; the field will then consist of a relatively small number of turns of thick wire.

The series motor is particularly suited to engine starting requirements, since it develops its maximum torque immediately the motor is switched on, i.e. when the engine starting, or breakaway, torque is at a maximum. As the engine speed increases from rest the series motor torque decreases; the motor current then falls off.

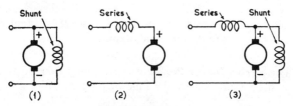

FIG. 10. Types of electric motor. (1) shunt. (2) series. (3) compound, or series-shunt

Since the main current is also the flux-producing or magnetizing current it follows that the induced electromotive force is proportional to the product of the speed and field strength. Thus:

$$\text{Speed of motor} = \text{constant} \times \frac{\text{induced e.m.f.}}{\text{field}}$$

Therefore, for a given field strength the e.m.f. induced varies directly as the speed. Further, if the voltage (battery) is constant the voltage drop in the motor resistance is so small that it can be omitted, so that

$$\text{Speed of motor} = \text{constant} \times \frac{\text{applied voltage}}{\text{field}}$$

The effect of additional load on the series motor is to reduce the speed and thus increase the field load. Reduction of load increases the speed so that under 'no-load' conditions, the speed may become dangerously high. The torque developed by a series motor is proportional to the product of the current and the field, and is independent of the motor speed. Thus:

Motor torque = constant × current × field

Under constant flux conditions the current value in a series motor determines the torque and the applied voltage, the motor speed.

Series motors are usually of the four-pole type, using single turn copper strip coils for the armature, inserted axially into tunnel-type slots. The pole windings, as stated earlier, are in series with those of the armature. If the armature coils are *lap-connected* there will be four brush spindles; if *wave-connected*, either two or four spindles. For 6-volt motors the former method of connection is employed, so that there is twice the current value as with the 12-volt motor of equal output.

(2) The Series Motor Performance. The power of a series motor can be expressed by the following relationship:

$$P = CV_R + C^2 R$$

where C = current, V_R = battery internal voltage and R = sum of the resistances of the battery, armature and brushes, the starter field coils and the leads. The power P is expressed in electrical units. The useful output for mechanical power is CV_R.

The maximum power is given by:

$$P \max. = \frac{V_R^2}{4R}$$

This expression shows how important it is to keep the resistances of the leads, i.e. connections between the battery, switch and motor as low as possible; also that of the battery, itself.

It also follows, from the same expression that if the voltage of the battery is reduced to one-half, then for the same motor power output it is necessary to reduce the resistance R to one-quarter; this can only be done by increasing the sectional areas of such items as the leads and brushes.

The substitution of the 12-volt for the 6-volt battery system in British and more recent American cars is an example of the saving

THE STARTING SYSTEM 39

in the copper of current conductors in starting motors and also in starting switches and leads.

The characteristic curves of a 12-volt starting motor are reproduced in Fig. 11, for a 4⅛ in. diameter motor operated over speeds up to 2,500 r.p.m. The motor, it will be observed, developed a maximum output of about 1·2 b.h.p. at 1,200 r.p.m. and maximum (locked) torque of just over 15 lb. ft. for starting purposes.

Assuming a starter-pinion to flywheel ring gear ratio of 1 : 15, and a cranking speed, for overcoming the engine resisting torque, of 50 r.p.m., the corresponding motor armature speed will be 750 r.p.m.

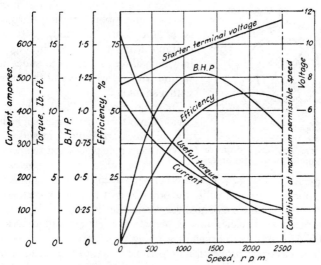

FIG. 11. Performance curves for starting motor

The resisting torque at this speed will be about 8 lb. ft. and the motor output about 1·14 b.h.p.

It is characteristic of automobile starting motors that under locked torque conditions, large currents are taken from the battery. Thus, in the present example a current of about 445 amps. is taken at the moment of switching on the starting motor, but this current falls to about 280 amps. during the cranking speed operation, mentioned earlier. It will be seen that the starting motor voltage, at starting, is about 7·6, increasing as the load is reduced, to about 9 volts at the cranking speed.

When the starting motor is developing its maximum power its terminal voltage is generally about 75 to 80 per cent of the open circuit battery voltage, i.e. about 9 to $9\frac{1}{2}$ volts.

(3) **The Compound, or Series-shunt Wound Motor.** In this type two or three of the field coils are connected in series, and these coils are connected in shunt with the armature windings, as shown in Fig. 10 (3).

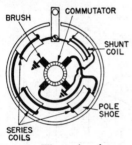

FIG. 12. The series-shunt or compound motor (Rambler)

This type of motor which is much used for automobile starting purposes is of the four-pole kind, a typical example being that shown in Fig. 12, in which three of the pole shoes are wound with series coils and the fourth with a shunt coil.

In some cases the field current is divided into two equal parts, such that two windings are applied, in series, to two adjacent pole pieces and two windings to the other pair of poles as depicted in Fig. 13 (a). The other ends of the coils are connected to opposite brushes, while the other pair of brushes is earthed as shown. Sometimes an equalizer connection is employed between the insulated brushes to compensate for current differences in the field coils due to brush or winding variations.

The electrical properties, or performances, of the series-shunt motor are intermediate between those of the series and shunt motors and may be enumerated, briefly, as follows:

(1) The motor field current is practically constant, so that the magnetic field is almost constant.
(2) The torque is proportional to the armature current.
(3) Small variations in the magnetic field due to armature current have a relatively small effect on the speed of the motor, e.g.

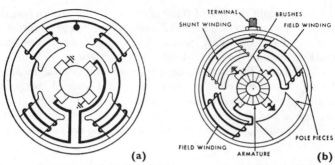

FIG. 13. (a) Motor with divided field current coils. (b) Four-pole motor, having one idler pole

usually, not more than about 10 per cent between no-load and full-load.

The series-shunt motor can be provided with any required combination of the properties of the series and shunt motors but, in general since the field strength increases with the current strength, the speed will fall as the load increases, so that the speed falls off as the torque is increased, while the torque increases as the current increases.

Three-Pole Motor. A type of series-shunt motor widely used on the American Ford, Mercury and Lincoln cars, employs series windings on oppositely-located poles with a shunt winding on one of the intermediate poles, as shown in Fig. 13 (b). The other pole is not utilized for any winding but the motor performance is adequate for its designed purpose.

Selection of Starting Motors. It will be apparent, from earlier considerations that since the starting and cranking torques of engines depend upon their sizes, the starting motor characteristics must be selected to match the starting requirements.

For each type and size of engine it is necessary to know the torques required to crank the engine over a range of possible starting speeds. It is, however, more convenient to plot the cranking horse-power against engine speed and to use this graph for determining the size of starting motor. In this connection the torques should be ascertained under low temperature, e.g. freezing point conditions, and, with regard to the motor account must be taken of the pinion-to-flywheel gear ratio.

To illustrate these points, the cranking power curves of a small Diesel engine of about 2-litre capacity and a small car petrol engine of 1-litre capacity are reproduced in Fig. 14 by the graphs marked A and B respectively. The power output curves of two starting motors, namely, C and D, are shown on the same diagram.

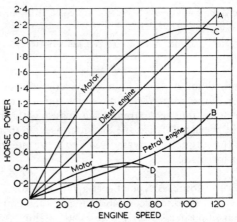

FIG. 14. Performance test results for Diesel and petrol engine starting systems

In the case of the Diesel engine the fact that the motor output graph lies well above the engine power graph indicates that the motor develops more power and is thus able to accelerate the engine until, where the two graphs A and C meet the two powers are equal; this shows that at the engine speed corresponding to the intersection of the graphs, namely, about 112 r.p.m. the maximum engine cranking speed is reached.

Similarly, for the petrol engine the starting motor will accelerate the engine up to a maximum cranking speed of 64 r.p.m., which is sufficient to enable the engine to start on its ignition system.

THE STARTING SYSTEM 43

The corresponding armature shaft speeds of the starting motors C and D, at their maximum cranking speeds, assuming a gear ratio of 1 : 10 will be 1,160 and 640 r.p.m., respectively. The motors in question were designed to operate off batteries of about 160 and 35 ampere-hour capacities, respectively.

It should here be mentioned that by varying the gear ratio between the motor and flywheel gears the torque values at any given speed can be altered so that any given motor's engine starting characteristics can be changed to a certain extent.

Some Starting Motor Data. The range of starting motor sizes employed for automobile and commercial vehicle purposes varies from about $3\frac{1}{2}$ in. diameter for the smaller 6- and 12-volt models used on light cars to about 6 in. diameter, operating on 24 volts, for larger commercial vehicles, fitted with Diesel engines.

The smaller 6-volt starting motors have a locked torque of about 5 to 7 lb. ft., while the 12-volt motors used for $1\frac{1}{2}$-litre engines have locked torques of about 8 to 10 lb. ft. and starting currents of 350 to 400 amps., on batteries of 40 to 45 amp. hr. (at the 20-hour discharge rate. Their light-running current values are usually about 45 amps.

A typical starting motor for a modern American car engine, of the 90° vee-eight type developing a maximum output of 325 b.h.p. at 4,400 r.p.m., would employ a battery of 70 amp. hr. at the 20-hour discharge rate.

The starting motor (Delco-Remy), has a gear ratio of 18·4 : 1 and will crank the engine for starting purposes up to 160 (engine) r.p.m. At normal air temperatures, the locked torque is from 13 to 15 lb. ft., with a current draw of 450 to 550 amps. and, for the 12-volt battery used, a corresponding voltage (lock-test) of 4·0 to 5·0.

For a six-cylinder Diesel engine, developing 130 b.h.p., a 24-volt, 100 to 120 amp. hr. battery would be used. In a typical example the flywheel gear-to-motor pinion gear ratio is 13 : 1. The motor is series-wound and has a locked torque of 38 lb. ft. and current draw of 600 amps. The maximum motor output is 3·1 h.p. and the torque at maximum output is 13·˙˙ lb. ft. The current taken at maximum output is 250 amps. and the motor speed is 1,200 r.p.m.

Starting Motor Efficiency. The mechanical output of an electric motor is always less than the mechanical equivalent of the electrical input of the motor, owing to friction, windage and electrical losses. The efficiency E, for an electrical input of W watts at the motor terminals and an output horse power, denoted by $H.P.$ is given by the relation:

$$E = \frac{H.P. \times 746}{W} \times 100$$

Here, $W = C \times V$, where $C =$ motor current and V the motor terminal voltage.

Usually, the starting motor efficiency lies between about 60 and 75 per cent, under the most favourable conditions.

The Commutator Brushes. In common with the starting motor which is designed for heavy intermittent duties, but not for continuous running the commutator brushes must be arranged to withstand the effects of high current flow per sq. in. of contact area. Thus, whereas for normal running purposes, electric motor brushes normally carry currents between 50 and 100 amps. per sq. in., those of starting motors are occasionally called upon to take lock-torque or starting current densities up to 1,000 to 2,000 amps. per sq. in.

In the case of a normal size of brush, namely $\frac{5}{8}$ in. $\times \frac{5}{16}$ in., which carries a starting current up to, say, 300 amps., the current density works out at 1,500 amps. per sq. in. For heavy duty car motor purposes metal-graphite brushes containing 70 to 80 per cent bronze are generally used, while for the smaller cars, still using 6-volt systems, carbon brushes are employed. The heaviest types of starting motor, as used for Diesel engine starters, operating on the 24-volt system, use metal graphite brushes of 50 to 70 per cent copper.

The voltage drop should not exceed 0·5 volt per brush, so that special care is necessary when bedding brushes on their commutator to insure the best possible contact.

Brush spring pressures are usually about 8 to 12 lb. per sq. in. The mica insulation between the commutator copper bars is undercut in some cases usually where the brushes would be worn away by the mica edges if not undercut, but when the brushes are sufficiently abrasive in composition, undercutting is unnecessary.

In regard to the *life of starting motor brushes*, good quality ones should provide from about 6,000 to 9,000 starts or, on an average use mileage basis, from 25,000 to 40,000 miles, before replacement becomes necessary.

Checking Starting Motor Specifications. It is usual to specify the performance of each motor when new, the principal tests being those for *Locked or Stall Torque* and *No Load*. Usually, the brush spring tension and solenoid (shift lever) pull-in and hold-in currents and voltages are also specified.

A typical Auto-Lite starting motor for cranking a six-cylinder engine of 130 b.h.p. has the following specifications:

Locked Torque. 6·5 lb. ft. (min.) with 285 amps. (max.) at 4·0 v.
No Load. 10 volts and 70 amps. (max.) at 5,300 r.p.m. (min.).
Brush spring tension. 40 to 50 ozs.
Solenoid (*Shift lever*). Pull-in Coil Draw. 19 to 23 amps. at 6 volts.
                               Hold-in Coil Draw. 10 to 12 amps. at 6 volts.

Locked Torque Test. The layout of the test equipment for Rambler starting motors is as shown in Fig. 15 (A). Here, the motor is clamped in a special vice and a lever, or torque arm is rigidly secured to the starter pinion. At a given distance from the motor axis—usually one foot—a spring balance is hooked on. A voltmeter is connected in shunt from the live terminal of the motor to the metal (earth) of the motor. The ammeter and a carbon pile rheostat are connected in series with a heavy duty switch and the battery. The carbon pile type of rheostat is used, since it can take large currents.

No Load Test. The layout of the equipment for this test is shown in Fig. 15 (B). The connections are similar to those for the locked torque test but no torque arm is employed. A tachometer

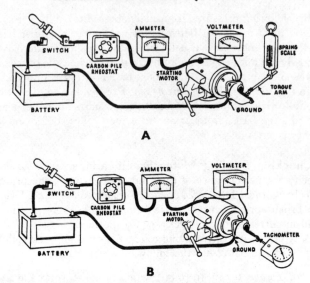

FIG. 15. (A) Making a locked torque test. (B) The no load test

is used to measure the motor speed. The current is increased slowly, after closing the switch by means of the rheostat so that the motor, now under free running, or no load conditions, will speed up to its maximum specified speed—usually from 4,000 to 5,500 r.p.m. The current and voltage under these high speed conditions are then read off and compared with the specified values.

Starting Motor Circuits. As mentioned earlier the electrical circuit of the starting motor—shown symbolically in Fig. 2—is the simplest of the five basic circuits comprising the complete electrical system. It consists of the battery, motor and special switching system, as shown schematically in Fig. 16.

When the starter button on the facia board is depressed the starter switch is closed. Current from the negative pole* of the battery energizes the solenoid switch plunger, causing it to move the relatively heavy switch plate to the closed position so that

* The positive pole in American systems.

THE STARTING SYSTEM

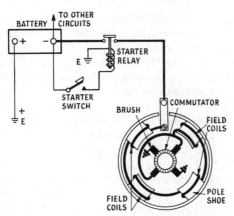

FIG. 16. Starting motor circuit, with solenoid switch, for four-pole series motor

current flows from the battery direct to the field coils of the motor. In this example a four-pole series type motor is illustrated. From the field coils the current flows through the commutator and armature windings and thence by way of the two positive earth brushes to the metal return path of the system.

In this diagram the heavy current circuit is indicated by the heavier black lines, and the low current value circuits by the finer lines. It will be seen that only a small current is required to operate the driver's facia panel switch and also, that none of the heavy current in the motor circuit passes through the ammeter or rest of the electrical system of the car or other vehicle. It will also be seen that one end of the solenoid coil is earthed.

In the case of cars having automatic transmissions a separate 'lock-out' switch is incorporated in the starter switching section, to prevent operation of the starter when the selector lever of the transmission is not placed in either the Neutral (N) or Parked (P) positions.

In some instances a driver's foot switch, designed to take the full starting ampere current, is employed and is then wired direct in series with the starting motor and battery.

The starting motor circuits hitherto described apply to motors which employ the Bendix or inertia-type of pinion gears, as described later. When the alternative type of starting motor with pre-engaged pinion gear is used, the circuit arrangement includes the windings of the special solenoid—mounted on the motor casing—which actuates the pinion engagement mechanism. In this system the usual solenoid starting switch is not required. Further reference to this system is made later.

Starting Motor Drives. Various kinds of starting motor drive are employed for cars and commercial vehicles, these being based upon the general principle of engaging the motor pinion (gear) with the flywheel gear, shortly after the starter switch has been operated and releasing the pinion from engagement as soon as the engine commences to operate under its own power. In some motor drives, the pinion is slid into gear with the flywheel gear while the motor is accelerating, while in other drives the pinion is first slid into gear, i.e. before the armature shaft commences to rotate and then the motor itself is switched on.

The principal starting motor drives used include: (1) The Bendix Pinion. (2) The Rubber Coupling Inertia Pinion. (3) The Manually-operated Over-running Clutch. (4) The Solenoid-operated Over-running Clutch. (5) The Axial or Sliding Armature.

(1) The Bendix Pinion. This has been the most widely used of the various starting motor drives on account of its simplicity and reliability, coupled with low production cost.

It belongs to the inertia-type of starter pinion, since there is a lag, due to the inertia of the pinion assembly, after switching the motor into circuit and before the actual engagement of the pinion with the flywheel gear.

The pinion is mounted on a hollow sleeve which has an external square thread to match the internal thread of the pinion; in practice the pinion thread is made a somewhat loose fit on the sleeve thread. The sleeve is splined at one end, the splines enabling it to slide on similar splines on the armature shaft. Alternatively, as in the original or standard Bendix drive one end of the sleeve is bolted to the Bendix drive compression-type spring (Fig. 17) and the other end of the drive spring is keyed

THE STARTING SYSTEM

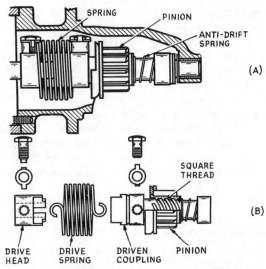

FIG. 17. Bendix pinion motor drive. (A) Complete assembly on motor armature shaft. (B) Components of Bendix drive

and bolted to the armature shaft through a drive head unit. When the engine is at rest the pinion is disengaged from the flywheel gear, but as soon as the motor is switched on the armature of the motor commences to rotate and quickly accelerates. The threaded sleeve, which is driven from the armature shaft through the drive spring also accelerates but the drive pinion, on account of its loose fit on the sleeve experiences an inertia effect, which causes it to lag, so that the sleeve rotates within the pinion, thus screwing the pinion axially along the shaft and into mesh with the flywheel gear. The pinion acts in a similar manner to that of a nut which is held against rotation for a time, while the screwed shaft member rotates, thus causing the nut to move in the endwise direction.

When the drive pinion meets the pinion stop at the end of its sleeve, since it can then move no farther, it is compelled to rotate with the sleeve, and it must accordingly drive the flywheel gear. The spring drive is compressed slightly at this stage, thus reducing the shock of the drive engagement.

After the engine begins to 'fire', the speed of rotation of the pinion exceeds that of the armature shaft, so that the pinion is released from mesh with the flywheel gear and returns to the disengaged position. This is an important phase since, should the pinion not be rejected, immediately after the engine 'fires' the armature will be accelerated to a dangerous speed. Thus, if the pinion is not released before the fast-idling speed of about 1,000 r.p.m. is reached by the engine, and the pinion-to-flywheel gear ratio is 15 : 1 the armature will be spun up to 15,000 r.p.m.

The Bendix drive is provided with a light compression spring to prevent the pinion teeth striking those of the flywheel gear while the engine is running; this spring is known as the *anti-drift spring*.

There are two kinds of Bendix drive namely (1) The Outboard and (2) The Inboard, according to whether the Bendix pinion slides *outwards*, i.e. away from the starting motor, for engagement

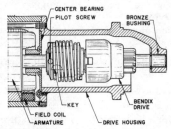

Fig. 18. The Outboard-type Bendix drive

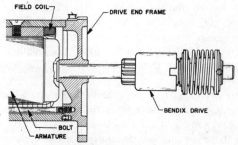

Fig. 19. The Inboard-type Bendix drive

with the flywheel gear, or *inwards*, i.e. towards the motor. These two types of drive are illustrated in the Delco-Remy applications, shown in Figs. 18 and 19. It will be observed that in the case of the outboard arrangement, an outside bearing is provided.

Various designs of Bendix drives are used, including inbuilt reduction gear and clutch types. The latter drive embodies a friction clutch between the armature and Bendix drive, which is spring-adjusted so as to slip momentarily to relieve the shock of engagement of the Bendix pinion and flywheel gear.

In the Bendix drive there is always a risk of damage to the flywheel gear teeth due to the starting switch being operated while either the engine or the motor is running. Further, after prolonged use of the Bendix starter the flywheel gear teeth become worn excessively at certain positions, corresponding to those in which the engine stops; these positions are also dependent upon the number of engine cylinders. Premature disengagement of the drive pinion is another possibility, while the starter pinion may occasionally become jammed in the 'off' position, so that the motor idles freely when switched on but does not rotate the flywheel. This fault is commonly due to dirt or grease between the pinion and sleeve threads.

Improved Inertia Pinion. Two mechanical disadvantages of the original Bendix pinion are (1) That this requires a relatively large pinion. (2) That the torsion spring becomes useless if one end fractures. These disadvantages have been overcome in the Lucas inertia pinion by making the splines in a sleeve attached to the (smaller) pinion and with the use of a compression instead of a torsion spring. This design is shown in the lower part of Fig. 30, which shows the various components, including the relatively small pinion and compression spring; if the coil fractures in one place the spring will still continue to function, i.e. this will not immobilize the starter. The inward meshing of the pinion is another advantage, this arrangement bringing the pinion with its heavy torque reaction nearer to the armature so as to give a shorter and stiffer shaft.

The Bendix Folo-Thru Drive. This improved Bendix drive is designed to overcome premature disengagement of the drive pinion and flywheel gear, until a given engine speed has been reached.

Fig. 20 shows this drive, as used on Chrysler car engines. In operation, the Bendix drive engages with the flywheel gear and rotates the crankshaft, but during this operation a spring-loaded detent pin, located in the control nut drops into engagement with a notch in the screwshaft. If the engine ceases to 'fire', movement

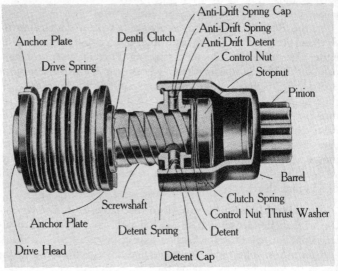

FIG. 20. The Bendix Folo-Thru drive

of the control nut in the disengaging direction on the screwshaft is checked, momentarily, because of the pin engagement in the screwshaft notch, so that the pinion cannot be released from the flywheel gear. When, subsequently, the engine fires regularly and its speed attains a predetermined value the detent pin is thrown out of its notch by centrifugal action, when the pinion will disengage from the flywheel gear, automatically, in the usual manner.

(2) **The Rubber Coupling Inertia Pinion.** This type of drive, while employing the inertia pinion engagement method, also

embodies a combined rubber drive and friction coupling to transmit the drive from the starter shaft to the pinion. Fig. 21 illustrates the Lucas drive which employs this principle. The rubber member is a rubber bush secured to an inner sleeve and pressed securely against an outer sleeve. The torque is transmitted from the motor partly by way of the rubber coupling and partly by means of a friction washer. This arrangement provides a normal cushion drive, due to the rubber acting in shear, and a

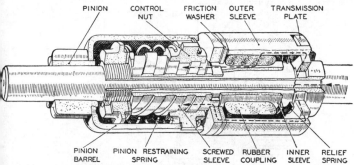

FIG. 21. Rubber coupling inertia pinion unit, showing paths taken by the driving torque

means of permitting slipping when the drive is much overloaded. By proportioning the parts so that slipping occurs between the rubber bush and the outer sleeve of the coupling unit, the torque at which slipping will take place can be varied to suit the particular type of engine. Under normal engagement conditions the rubber acts merely as a spring with a limiting relative twist between the inner and outer sleeves, of about 30°. When severely overloaded, such as may occur due to a backfire or to the pinion being inadvertently slipped into mesh when the flywheel is rotating in a reverse direction, the rubber slips in its housing. Thus no damage can occur to the starting motor drive or its mounting when overloaded.

The pinion restraining spring, shown in Fig. 21, is fitted between the rubber coupling unit and transmission plate, to take the shock if the teeth meet end to end, thus allowing the pinion to get into mesh with the flywheel teeth.

(3) The Manually-operated Over-running Clutch. In this type of starting motor drive the heavy duty switch is mounted on top of the motor casing and is operated by a cable-controlled lever—usually actuated by the driver's pedal starter—which also moves the pinion unit, axially along the armature shaft, into engagement with the flywheel gear. The pinion can be disengaged, after releasing the 'Shift Lever' (Fig. 22) by means of

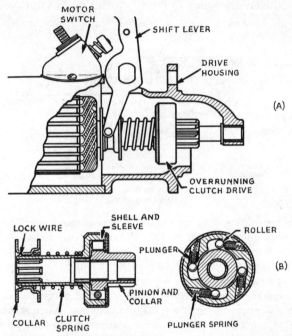

FIG. 22. Manually-operated over-running clutch drive (Delco-Remy). (A) Complete assembly. (B) Details of clutch unit

the compression spring located between the lower end of the shift lever and the pinion-clutch unit. After the engine has started, the over-running clutch permits the drive pinion to over-run, so that if there is any delay in operating the shift lever, no damage is done to the pinion or armature unit and the latter unit is thus

THE STARTING SYSTEM 55

protected from excessive speed during the short interval that the drive pinion remains in mesh.

The Delco-Remy *over-running clutch* consists of a shell-and-sleeve assembly which is splined to fit the armature shaft splines. A pinion gear and collar unit fits loosely into the shell and the collar is in contact with four hardened steel rollers which are assembled into notches cut in the inner face of the shell (Fig. 22 (B)). These notches taper inward slightly so that there is less room at the rear end. The rollers are spring-loaded by small plungers. When the pinion is moved towards the flywheel gear, if the teeth should butt on those of the flywheel gear the clutch spring compresses so that the pinion is spring-loaded against the flywheel gear teeth. Then, when the armature begins to rotate meshing takes place at once. Completion of the shift lever movement closes the motor switch and the armature starts to rotate. This rotates the shell-and-sleeve assembly, thus causing the rollers to jam tightly in the smaller sections of the notches. The rollers actually jam between the pinion collar and shell so that the pinion is forced to rotate with the armature and therefore will then crank the engine.

The same kind of over-running clutch is used with the solenoid operated shift lever, next to be described.

(4) The Solenoid-operated Over-running Clutch. In the case of larger car and vehicle petrol engines and also, invariably, for Diesel engines, the starter pinion is first moved into mesh, automatically with the flywheel gear; then the starter switch is operated. Afterwards, the motor is switched on and the already-engaged pinion at once begins to rotate the flywheel.

In this connection the principle is the same as for the previously described manually-operated shift lever. Before describing this type of drive it is proposed to mention the subject of Diesel engine starting.

Diesel Engine Starting. While under normal temperature conditions a Diesel engine will usually commence to operate under its own power, at once, this is not usually the case under freezing point temperature conditions. It is then necessary to motor the engine for an appreciably longer period than for petrol engines. The acceleration of the Diesel engine flywheel is such that at the first firing stroke the inertia type of starter pinion would be thrown out of engagement prematurely and in most cases the engine would stop. It is therefore necessary to provide the alternative to the inertia pinion, namely, the pre-engaged pinion so that the engine can be motored until it starts and then begins to run regularly.

Solenoid-operated Pinion Drive. This widely-used starting motor engagement device employs a special kind of solenoid switch mounted on top of the motor. When the driver's hand switch is pressed, battery current of low amperage flows around the solenoid draw-in coil (Fig. 23) and causes a plunger to move

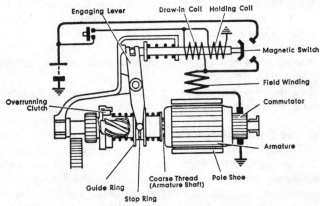

FIG. 23. Solenoid-operated over-running clutch drive principle

to the right. This plunger movement first causes a pivoted engagement lever to move the motor drive pinion into engagement with the flywheel gear, after which further movement of the plunger causes the magnetic switch, on the right, to close so that the motor field and armature circuits are energized. The pinion which is already engaged with the flywheel teeth, rotates the flywheel, all the time the driver's switch is engaged. When this switch is released the solenoid plunger returns to its original position, bringing with it the end of the engaging lever, thus withdrawing the pinion from the flywheel gear. An over-running clutch must be used with this kind of starter; the type used on the Bosch 'screw-push' starter being similar to that illustrated in Fig. 22.

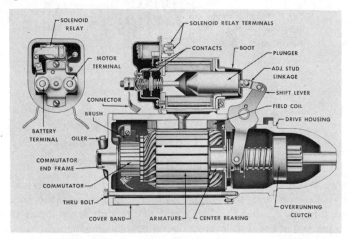

Fig. 24. The Delco-Remy solenoid-operated over-running clutch drive starting motor

Fig. 24 shows the Delco-Remy solenoid-operated starter with its over-running clutch. The various components are indicated by lettering which should render the previously given explanation clear.

Fig. 25 shows the internal windings of the solenoid unit. There are two windings on the solenoid, namely, a *pull-in* or main current winding for providing the magnetic field to actuate the solenoid plunger and a *hold-in* winding. Both windings produce a magnetic field, but when the plunger moves sufficiently to close the main contacts—shown on the right in Fig. 25—the pull-in winding is short-circuited—since this winding is connected across the main contacts—and the magnetism produced by the hold-in winding is sufficient to hold the plunger in but with a much smaller current from the battery than that of the pull-in

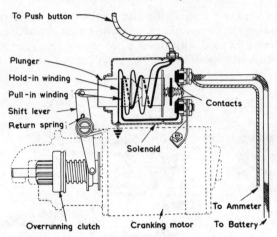

Fig. 25. Showing internal windings of solenoid unit

coil. When the driver's switch is released the hold-in current is switched off and the compression spring between the over-running clutch and the lower end of the shift lever forces the upper end of the shift lever to the right, thus drawing the plunger to the right and breaking the motor main current circuit, by opening the contacts shown on the right hand side.

In the solenoid switch arrangement shown in Fig. 24 it will be seen that there is a small relay unit mounted above the main motor contacts, on the left hand side of the plunger. In effect the relay is an additional switch which is operated by some external control circuit, but the operation of the two-winding type solenoid is the same as for the direct or driver's switch method of operation.

The over-running clutch is similar to that described in the previous section and its purpose is to prevent the engine from driving the motor at excessive speeds before the starter pinion is disengaged.

While this form of starter drive is more positive in action and is not liable to premature ejection of the pinion it is subject to the same kind of damage as the inertia type starter, as when attempting to engage the drive with the engine running.

Since the plunger would be liable to damage should water or

dirt enter the unit, the rubber boot, shown in Fig. 24, is provided for protection purposes.

(5) The Axial Sliding Armature Starter. The inertia-type pinion starter, while quite satisfactory for smaller to larger sizes of car engines, is not suitable for heavier vehicles, owing to the large alternating stresses encountered during the cranking of larger petrol engines and, in particular, Diesel engines. Further, as mentioned earlier, a single initial firing stroke of a Diesel engine would cause the pinion to be thrown out of its engagement with the flywheel gear.

For such transport engines in this country starting motors, in some cases exceeding 7 h.p. have to be used, with engine breakaway torques of 70 to 100 lb. ft. The initial locked torque current may rise to 1,000 amps. while the motoring or cranking speed current of 500 amps. may be needed, to rotate the crankshaft at these speeds, namely, 120 to 150 r.p.m. (engine speed).

The axial starter operates on the principle of a sliding armature mechanism, such that when the starter switch is operated a solenoid coil is energized. This brings into circuit a shunt winding and also an auxiliary series field winding. Due to the magnetic field thus excited the armature, as it commences to rotate slowly, is drawn axially into a central position. Since the pinion gear is attached to the armature shaft the pinion is moved axially and into engagement with the flywheel gear. During this operation the armature speed is limited by the shunt and auxiliary series field windings, so that the gears will mesh smoothly. After full engagement of the gears a catch on the switch is tripped, and this energizes the main series winding, so that the motor now operates on its full power current supply. As with the previously described starter, a slipping clutch is provided between the pinion and armature, to prevent overloading or overspeeding of the motor.

In order to prevent the engine from stalling after it has fired, the shunt winding and starter spring are designed so as to prevent disengagement of the pinion before the starter switch is released.

The axial starter can be made of smaller overall dimensions

than the solenoid shift lever type since the solenoid coils are arranged within the motor casing.

The Bosch Sliding Armature Starter. This starter employs an auxiliary winding and also a holding coil in addition to the main field current winding. When the driver's switch is operated it actuates a magnetic switch which has two consecutive actions, namely, (1) (Fig. 26 (A)) The armature is pulled into the exciting field of the auxiliary winding, so that a rather longer commutator

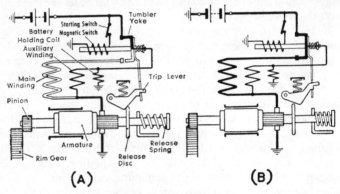

Fig. 26. The Bosch sliding armature starter drive. (A) Initial position of armature. (B) Engaged position

than usual is required. At the same time the armature is rotated slowly until the pinion teeth engage with those on the flywheel gear. (2) While only the auxiliary field winding and holding coil are 'live' during the first action, current now flows through the series winding because of the tumbler yoke which connects the battery and main current winding (Fig. 26 (B)). This is done by the release of the tumbler yoke which is locked by the trip lever during operation (1). The release disc travels along with the advancing armature and lifts the trip lever, in meshing the pinion thus allowing the tumbler yoke to make contact at its other end, also. As soon as the series winding is switched on the starter commences to rotate the crankshaft, and then accelerate it, thus reducing the armature current and therefore the pull on the armature. To prevent the pinion demeshing a separate holding coil in a shunt circuit is employed. When the driver releases the starter switch button the holding coil becomes de-energized and the pinion then disengages from the flywheel gear.

It is necessary to insert, as a resilient member, between the armature shaft and the pinion a multi-disc spring-loaded clutch in order to provide a smooth frictional connection. The release spring which pushes the armature shaft back to its initial position is located within the armature shaft (Fig. 27), and a recess at its

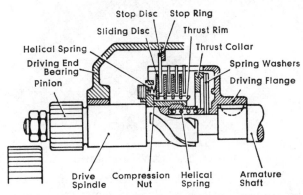

FIG. 27. The resilient multiple-disc clutch unit. *Note.* The lower helical spring is the release one

circumference assists the sliding of the armature. When the flywheel is suddenly accelerated, by the engine firing impulses, the clutches act by free-wheeling. The multi-disc clutch, in addition, acts as an overload protection against excessive torque.

The Co-axial Starter. This starter performs the same duties as the axial starter but it embodies certain special design features, including (1) Engagement of the pinion occurs under reduced power thus avoiding heavy engagement shocks and gear teeth wear. (2) The pinion *only* moves forward to engage the flywheel gear, instead of the whole armature as in the axial starter.

In this starter (Fig. 28) the pinion is moved forward into

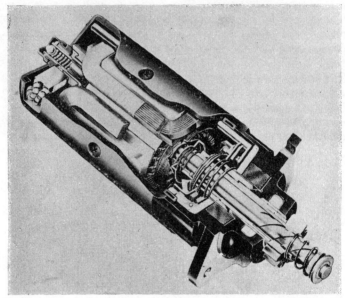

Fig. 28. The C.A.V. co-axial starter

engagement by a solenoid mounted co-axially with the armature, so that a smooth cylindrical shape of motor can be used. The starter is provided with a mechanical locking device which locks the pinion firmly in mesh until the starter switch is released. Special provision is made also for preventing mis-engagement of the pinion when the flywheel and pinion teeth are badly worn.

The solenoid actuates a hollow plunger to push the pinion axially into gear with the flywheel teeth. At the same time the movement of the plunger closes a pair of contacts mounted in the solenoid; this connects the battery to the field windings by way of a resistance incorporated in the starter. The resistance reduces the voltage applied to the windings so that the armature rotates at a low speed. The pinion, which then has a rotary and axial movement, is drawn into full mesh with the flywheel gear by a helix machined on the armature shaft. Just before the fully-engaged position is reached, a collar carried on the end of

THE STARTING SYSTEM 63

the pinion sleeves trips a trigger on the solenoid. This causes a second pair of contacts to close and short circuit the resistance, so that the full battery voltage is applied to the windings. The pinion is held in the flywheel gear engaged position by a mechanical locking device consisting of four steel balls located in holes in the pinion sleeve (as in Fig. 22 (B)). When the starter button is released, the plunger in returning to its initial position pushes back a collar, releases the balls and leaves the pinion free.

Carburettor Type Starting Device. This method, which is used in recent Buick cars, for starting the engine by depressing the accelerator pedal, utilizes manifold vacuum to switch off the starting motor once the engine starts to 'fire'. It employs a special switch built into the throttle body of the carburettor in a position suitable for its operation by the accelerator mechanism. The switch (Fig. 29) has a stainless-steel ball, a plunger, guide block, W-shaped contact spring and return spring mounted in a passage in the body flange, which is closed by a terminal cap containing the solenoid switch contacts for the starting motor. When the engine is at rest, with the throttle closed the ball rests

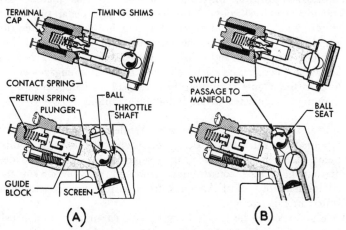

FIG. 29. Carburettor vacuum-operated starting motor switch
(A) Engine at rest. (B) Accelerator depressed.

on a lip on the lower end of the switch plunger and bears on a flat spot on the throttle shaft. The plunger, guide block and contact spring are held in a down position by the return spring, so that the contact spring does not touch the contacts in the terminal cap.

When the accelerator is depressed with the engine stopped and the ignition switched on the flat spot on the throttle shaft acts as a cam to push the switch ball, plunger, guide block and contact spring upward until the contact spring touches both contacts (shown by the black lines) in the terminal cap, thus switching on the motor starting current.

After the engine commences to 'fire' and the throttle is returned to the idle position the manifold vacuum forces the ball upwards against a seating in the throttle body, thus allowing the switch return spring to push the plunger assembly down and thus break the starting current circuit. The ball remains in this 'off' position all the time the engine is operating.

Very careful adjustment of the switch contact timing with the prescribed throttle opening is necessary to ensure correct starting conditions. Timing shims are provided for this purpose.

Semi-automatic Starting Scheme. In this method the starting switch is coupled to the accelerator pedal through a clutch, so that no hand-operated switch is required. After the engine has 'fired' the momentary releasing of the pedal disengages the clutch and allows the pedal to act as a normal accelerator pedal. The starter clutch is kept out of engagement as long as the engine is running by means of a vacuum-controlled diaphragm connected to the induction manifold.

Starting Motor Components. While this volume is not intended to deal with the design of automobile electrical equipment, it may be of interest to the reader to obtain a general idea of starting motor construction, from an examination of the components of typical examples, from current practice, as illustrated in Figs. 30 and 31.

Fig. 30 shows the Lucas Type M 35 G1 four-pole, four-brush series-parallel starting motor, fitted with Lucas inertia-type pinion and controlled by a solenoid switch mounted on the

THE STARTING SYSTEM 65

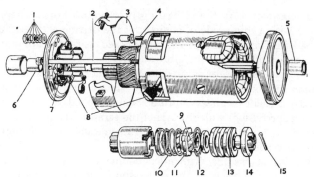

Fig. 30. The Lucas type M35G.1 four-pole series-shunt starting motor, and rubber coupling pinion device

1. Terminal nuts and washers
2. Through-bolt
3. Cover band
4. Terminal post
5. Bearing bush
6. Bearing bush
7. Brush spring
8. Brushes
9. Sleeve
10. Restraining spring
11. Control nut
12. Retaining ring
13. Main spring
14. Shaft nut
15. Cotter pin

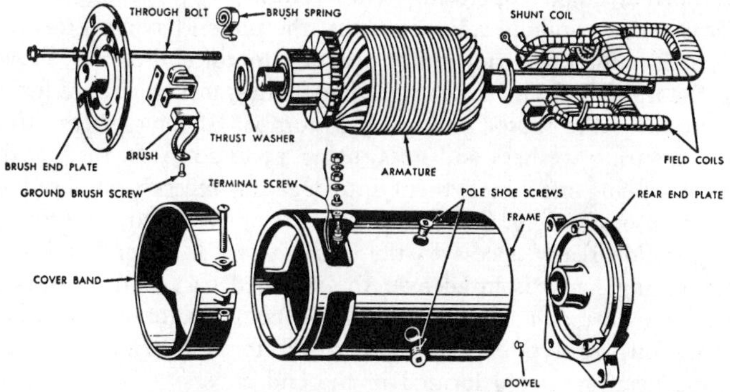

Fig. 31. Components of starting motor used on American Ford Company's cars

facia board and operated by movement of the ignition switch—as in most American cars—from a 12-volt battery.

The type used for 1·5 litre engines has a lock torque of 9·0–9·5 lb. ft. at 370 to 390 amps., at 7·7 to 7·3 volts. The light or no-load current is 45 amps.

The armature shaft end bearing is of porous bronze, impregnated with a thin grade of engine oil; the driving end bearing is a ball bearing, of the grease pre-packed self-lubricating type, requiring no maintenance lubrication.

The pole shoes are secured to the cylindrical housing of the motor by countersunk screws with slotted or recessed square heads, a relatively heavy torque being applied to each screw, with a special tool which prevents the screwdriver from slipping out of the screw slots or recesses. In some instances an impact-type driver is employed.

Fig. 31 shows the dismantled components of the American Ford starter. This is of the series-shunt wound type with two series and one shunt winding, four-pole, four-brush design, fitted with a Bendix or Folo-Thru pinion drive.

The starter gives a normal cranking (engine) speed of 150–180 r.p.m. for the 240 h.p. Interceptor vee-eight engine. The minimum torque is 15·5 lb. ft. at 5 volts, with a current draw of 550 amps. The no-load current is 80 amps. and the current draw at normal engine operating temperatures, 155–190 amps. The flywheel gear–pinion ratio is 16·2:1, the pinion having 9 teeth.

As is common starter motor practice, motor end covers, shown at the upper left and lower right hand sides, in Fig. 31, are firmly secured to the motor casing by a number of long or 'through' bolts, spring washers and nuts, using 15 to 20 lb. ft. torque, the covers being machined to fit into locating recesses in the ends of the motor casing. One end of this casing is slotted, generously, to provide ready access to the brush gear. A cover band with tightening screw is fitted over the recessed part of the casing to exclude dust and moisture from the commutator and brushes. Plain oil-less type bearings are used for the armature shaft, these bearings being located in the end covers.

CHAPTER 3

THE CHARGING SYSTEM

THE automobile's charging system has to provide electrical energy for all the load demands at the proper voltage, while preventing excessive voltages at the maximum engine speeds.

In recent years the amount of electrical equipment on the automobile has increased rapidly, with the result that there is now a wide range of loads to be provided for by the charging systems.

The dynamo, or generator, which supplies the electrical energy, *via* the battery, must be capable of operating efficiently over a considerable range of speeds. In this connection, in order to maintain a lead-acid battery at the correct terminal voltage it is necessary for the dynamo charging voltage for a single battery cell to vary from 1·8 volts, in the discharged state, to 2·7 volts in the fully-charged state; this is a variation of 50 per cent. On the other hand, the dynamo, which is driven positively from the engine, must operate satisfactorily over a range of dynamo speeds, from about 600 to 7,000 r.p.m., i.e. with a variation of about 1,170 per cent. Two other important conditions, which necessitate automatic control in the charging system must be fulfilled, namely, as follows:

(1) The dynamo, which starts from zero speed, should be connected to the battery only when its speed is such that the induced electromotive force is equal to or above that of the battery voltage. Further, the dynamo should be disconnected from the battery, immediately its voltage drops below that of the battery. The automatic device used for this purpose is known as the *Cut-out* or *Reverse-current Relay*.
(2) Within its wide speed range the dynamo output must be regulated, by automatic means, so that the voltage is maintained at the value required by the battery condition. For

example, in a specific example of a car in which the dynamo cuts in at 16 m.p.h. and the maximum speed of the car is 80 m.p.h., the voltage of the dynamo must be regulated, correctly, over a speed range of 5:1. With the modern high performance production cars, which can attain speeds of 120 m.p.h., this range becomes 7·5:1. This regulation of voltage is carried out by an electrical device known as a *Voltage Regulator*, although over a long period a special design of dynamo of the *three-brush* kind without this regulator was employed for this purpose. Reference to this type of dynamo is made later in this chapter.

There is another kind of regulator which is standard equipment on American cars, but is becoming used to an increasing extent on British cars. It is known as the *Current-and-Voltage Regulator*, and is described later.

It should here be mentioned that in our present considerations only the direct-current type of dynamo is dealt with. Later, the more recent alternating current generator will be considered.

The Ideal Regulator. The ideal device would be that employing a regulator which would maintain the dynamo voltage constant, irrespective of the electrical load. In this connection the early Tyrill regulator would appear to meet these requirements. Basically, this regulator contained an electromagnet which was energized by dynamo current and attracted an armature which was provided with a pair of contacts to cut 'out' or 'in' the dynamo field. The effect of increased dynamo speed was to prolong the length of time the contacts were open, and thus reduce any excess voltage. In practice, so far as automobile requirements are concerned, several modifications to this comparatively simple method, become necessary, e.g. the insertion of resistances, rectifier, temperature compensator, spark erosion preventor, etc., and thus render it impractical from the viewpoint of complication, bulk and expense.

The Cut-out. Known also as the *Reverse Current Relay*, this device, as stated earlier, connects and disconnects the dynamo

THE CHARGING SYSTEM 69

to the battery according to the voltage of the dynamo. When the dynamo voltage is below that of the battery, the cut-out breaks the charging circuit, thus preventing heavy current discharge from the battery through the dynamo coils at low speeds and also when the engine stops.

The cut-out, for a 12-volt system, is usually arranged to close its contacts in order to complete the charging circuit at a voltage of about 12·7 to 13·3—this is known as the *Cut-in Voltage*.

Here, it should be stated that no matter what method of dynamo output regulation is employed a cut-out is always essential for direct current machines.

Principle of the Cut-out. Fig. 32 illustrates, schematically, the principle of the cut-out as applied to the shunt-type of

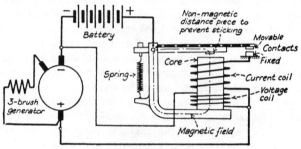

FIG. 32. The automatic cut-out, showing connections to battery and dynamo

dynamo which is the essential type for battery charging; in this example the three-brush dynamo has been chosen; further reference to this type is made later.

The cut-out unit consists, essentially, of a soft iron core around which are wound two coils, namely, (1) a fine wire *voltage* or *shunt coil* and (2) a thick wire *current* or *series* coil. Above the top of the core is a light iron arm, hinged at one end and provided with a contact, located above a lower fixed contact. Normally, a light spring holds the arm with the contacts open. When the contacts are closed there is a continuous magnetic field, as indicated by the looped dotted line, in Fig. 32.

70 MODERN ELECTRICAL EQUIPMENT

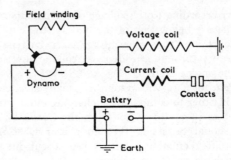

Fig. 33. Conventional diagram for the cut-out circuit, with three-brush dynamo

Assuming that the engine has been started and is running at its slow idling speed, the dynamo will be driven slowly and will produce a low voltage at the terminal coils, so that a small value current will flow around the shunt-coil circuit, to produce a low strength magnetic field. If, now the engine speed is increased the magnetic field will become stronger until at a dynamo speed usually of about 1,000 to 1,200 r.p.m., the field becomes strong enough to cause the light arm to be attracted to the soft iron core, or pole. This movement of the arm closes the contacts on the right and therefore switches the series coil into circuit with the generator and battery, so that the magnetic pull on the cut-out arm (and its contact) increases and thus the contacts are held together more tightly. The effect of closing the contacts does not affect the terminal voltage, since this is now controlled by the battery and regulated by the third brush of the dynamo. The value of the dynamo voltage at which the cut-out contacts close is usually between 12·7 and 13·3 volts.

When the engine speed falls to such a value that the E.M.F. in the generator is below the battery voltage, then the current reverses and a discharge current from the battery flows in a reverse direction to the charging current so that its magnetic effect opposes that of the current in the voltage coil, so that the magnetic pull on the lever above the core is reduced, finally allowing the control spring to open the contacts. The tension of the spring controls the value of the current, so that when the contacts close it is about 0·5 amp.; when the contacts open the current has a rather greater value. It is usual to insert a non-magnetic disc between the movable arm and the end of the core to prevent any sticking effect, unless a small air gap is provided for in the design.

The cut-out is essential in connection with the dynamo, battery and the regulator unit, irrespective of the kind of regulator used, and it works separately from the regulator although it is usually mounted on the same base as the regulator and enclosed in the same cover that protects the regulator's coils and contacts. Examples of these combined regulator and cut-out units are referred to again later in this chapter.

In many *commercial vehicles*, due to heavier currents that have to be controlled, the cut-outs are provided with *main and auxiliary contacts*. The latter contacts are arranged to close first and to open after the main contacts. In this way any sparking effects are avoided at the main contacts.

Usually, once the cut-out contacts are adjusted, little trouble is experienced, but it is possible to *adjust most cut-outs* for the correct closing voltages, namely, 12·7 to 13·3 volts for a 12-volt battery, 6·3 to 6·6 volts for a 6-volt battery and 25 to 27 volts for a 24-volt battery; in most cases an adjusting screw, having a locknut, is provided for varying the tension of the armature lever spring. The gap between the contacts, which is ·025–·030 in. for Lucas equipment, is adjusted by bending one of the arm or blade members.

The Dynamo. Before passing on to the modern two-brush type of dynamo, a brief reference will be made to the once popular three-brush dynamo, since it has one or two features of special technical interest. Referring to Fig. 32 it will be seen that in addition to the two main brushes, situated opposite one another on the commutator, there is a (smaller) third brush located a short distance from one of the main brushes. The shunt field winding is connected between the third brush and the other, more distant main brush. The electrical load, e.g. the ignition, lamps and certain other accessories is connected across the main brushes when the automatic cut-out is closed. The effect of the third brush may be understood by reference to Fig. 34, which shows the potential curve of the commutator under no-load conditions. The positions of the two main brushes on the 'straightened-out' commutator base are shown at A and B, while that of the third brush is indicated at C. The voltage between the main brush A and third brush C is proportional to the shaded area of the magnetic field distribution curve. The field-exciting winding is connected between A and C.

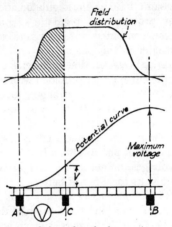

Fig. 34. Potential curve of three-brush dynamo commutator at no load

It can be shown that if the main field strength and dynamo speed are kept constant the voltage across A and C decreases rapidly as the current increases. This is due to the fact that the armature reaction is much more effective between the brushes A and C than between the main brushes A and B. In this way, use is made of the armature reaction to maintain a relatively constant voltage at various dynamo speeds.

If the dynamo is operated at constant speed while the electrical load is increased, gradually, the generator terminal voltage falls off. When, however, the generator and battery are connected in parallel, the dynamo voltage characteristic curve (on a current base) is modified so that the battery and dynamo have a common voltage, the value of which is determined by the battery condition and independently of the engine speed. In effect, the battery controls the voltage while the third brush dynamo controls the current output.

Varying the Current Output. The output of the dynamo can be altered between certain limits by altering the position of the third brush in relation to the fixed brushes, i.e. by varying the distance AC, in Fig. 34.

With the brush in its *normal* position, in a typical example the maximum output current was 5·6 amps. and the maximum voltage 15, at 2,500 r.p.m.

When the third brush was moved *backward* by a small amount the maximum current was reduced to about 4·8 amps. at 12·5 volts, at 2,500 r.p.m.

When the third brush was moved *forward*, the maximum current and voltage, at 2,500 r.p.m. were 10·5 amps. and 17·5 volts, respectively. Usually, back movement of the third brush by about a segment on the commutator will halve the normal current value.

Altering the Charging Rate. When the third brush dynamo was used, almost universally here and in the U.S.A. for automobiles, the charging rate was often varied by means of a three-position switch which provided: (1) The normal field resistance in shunt between the third brush and one main brush. (2) The insertion of an additional resistance in the dynamo field circuit, and (3) The insertion of a further resistance. This gave the choice of three maximum charging rates. Thus, in a typical British 12-volt system three maximum charging rates of 12, 10 and 7·5 amps. could be obtained, as indicated in Fig. 35.

Disadvantages of the Third Brush System. While this system represented an important step in automobile development, up to about 1933–36, its drawbacks were well known and subsequently this method was dropped and replaced by the constant voltage control method, described later. The chief disadvantages of the third brush system were as follows:

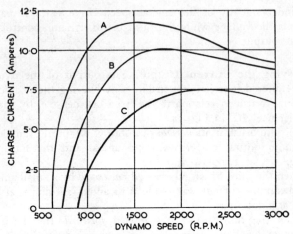

Fig. 35. Regulation curves of three-brush dynamo. (A) Without extra resistance in field circuit. (B) With single extra resistance. (C) With double extra resistance

(1) It produced its higher current values when the battery was charged and its lower values when the battery was in the partly-charged to low-charge condition, during daylight driving conditions, when no lamps were switched on. There was thus the tendency to overcharge the battery under daylight conditions. Hence the use of the three-position field resistance switch, previously mentioned.

(2) The current rises to a maximum as the dynamo speed is increased (Fig. 35), this maximum occurring at about 50 to 70 per cent of the maximum speed, thus there is a falling off in charging current at higher road speeds, so that part of the electrical load will be taken from the battery; further, when the battery is in a low state the more likely is this to occur.

(3) The insertion of additional resistances into the field circuit, in order to reduce the charging current, under daylight driving conditions, also raises the cutting-in speed of the cut-out.

THE CHARGING SYSTEM

Overcoming the Disadvantages. In order to obviate or reduce the effects of these disadvantages various schemes were put forward. These included the driver's switch for controlling the charging rate, the use of battery condition indicators on the dashboard, to enable the charging rate switch to be operated correctly; the overcharge prevention device, consisting of a relay which would reduce the charging rate when the voltage became too high and *the thermally-operated switch* or contacts, to insert a resistance into the field circuit at a given temperature. The principle of this method depends upon the fact that the temperature of the dynamo is a function of its output, so that as the temperature rises the thermostat will close a pair of contacts at some predetermined value, to insert the additional field resistance. While this device prevented excessive charging currents to the battery at a high voltage under daylight conditions it had the drawback of dependence upon atmospheric temperature conditions and dynamo location in regard to the radiator cooling air stream.

Importance of Battery Connection. Since the battery forms part of the third-brush system and determines its voltage, it is important *always* to use a battery when testing this type of dynamo. Otherwise, at the higher dynamo speeds excessive voltages (and currents) will be developed which may endanger the windings or any lamps used for loading purposes. Further, a cut-out switch for isolating the battery should never be used when testing this dynamo.

The Two Brush Dynamo. The standard design of modern automobile charging dynamos is the two-pole, two-brush model of the shunt type, i.e. with the field winding shunted across the armature winding at the brushes (Fig. 36). The shunt dynamo is the only one of the three available kinds, namely, series, compound and shunt, that is suitable for battery charging purposes, since this is the only type in which the polarity of the dynamo remains constant, irrespectively of whether the battery is charging or discharging. Further, by the insertion of an additional resistance in the field circuit the exciting current can be kept constant or varied, as required.

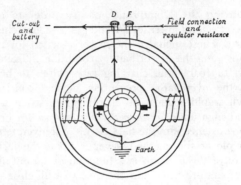

FIG. 36. The two-brush shunt-wound dynamo

Construction. The construction of a typical British automobile dynamo is illustrated in Fig 37, for the model used on $1\frac{1}{2}$ to 2-litre cars. Of relatively small dimensions, this dynamo is of the 12-volt type, having a maximum output of 230 watts, and

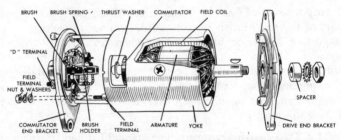

FIG. 37. Typical automobile dynamo

maximum current of 19 amps. The cut-out operates at a dynamo speed of 1,050–1,250 r.p.m. The maximum current speed is about 2,000 r.p.m. and the corresponding voltage 13·5. The brush-spring pressure is 22–25 oz.

Fig. 38 shows, in exploded view, all of the components, with their nomenclature, for the larger model Lucas C40–1 dynamo, as used on larger cars and certain commercial vehicles.

The yoke or shell is of welded, bent steel strip, being of final cylindrical form; alternatively the shell is made by the extrusion

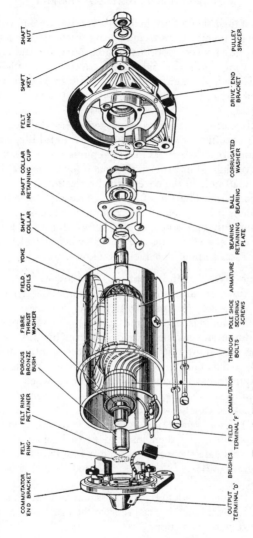

FIG. 38. The Lucas C40-1 ventilated shunt-wound dynamo, in exploded view

process, as a final shell. The shell is held to its two end covers by long 'through' bolts with nuts and spring washers. The left-hand cover (Fig. 38) houses a porous bronze bush for the armature spindle bearing while the right-hand cover has a ball-bearing. Provision is made for a small amount of end movement to allow for thermal expansion effects. The left end cover houses the brush gear.

The ball-bearing is of the pre-packed grease-lubricated kind and requires no further attention during its useful life, while the porous bronze bush is soaked in light engine oil before fitting to the end cover hole, but provision is made for occasional lubrication by a small hole in the cover; this lubrication is usually done every 6,000 miles. The life of the ball-bearing is about 50,000 miles, which is roughly that of the commutator brushes. It should be mentioned that in some of the larger dynamos a ball- or roller-bearing is used in the end cover; when a ball-bearing is used some axial sliding movement is allowed for.

Charging Dynamo Information. Two low-carbon steel pole pieces are fitted in the usual sizes of automobile dynamo but four-pole pieces are often used in larger vehicle sizes.

The field windings of dynamos are usually of copper wire, although aluminium has been used in Germany and elsewhere; the copper wire, however, renders the process of joint-making easier and, incidentally, provides a construction giving a rather higher output from the dynamo. The field coil, when completed, is taped and impregnated with a waterproof insulating compound.

The armature can be wound by one of three chief methods, namely, (1) *Form wound*, in which the coils are wound on a former, separately, and then inserted in their slots. (2) *Shuttle Wound*. In this case there is a separate winding bobbin for each coil. (3) *Machine Wound*. There each armature coil is wound, in turn, in place.

Due to the better strength and abrasion-resisting qualities of modern enamels this coating can replace the usual cotton covering for armature wire insulation.

The Dynamo Commutator. This unit is usually of the orthodox copper segment, mica or resin-bonded, strip insulation

THE CHARGING SYSTEM 79

kind, but a lighter construction is now possible with the moulded type of commutator.

The fact that the dynamo has to operate over such a wide speed range indicates that there must be a marked field distortion at the pole tip, when high speeds are attained. Such a distortion leads to excessive losses in the armature and at the commutator, due to faulty commutation. This, in turn, results in an overheating tendency in the commutator and this may cause melting of the soldered joints, should the temperature reach 70° C. Further possible ill effects are charring of the insulation and heating of the end bearing.

It is necessary, therefore, to provide *special internal cooling means* for the armature and commutator. In modern dynamos, a ventilation system is provided by the use of an extractor fan—which is usually integral with the driving pulley—and liberal air flow spaces at both ends of the dynamo.

An example of a modern well-ventilated dynamo is that of the Bosch Type LJ/GTL, shown in Fig. 39. This belongs to the larger output, namely 500–1,000 watt, class operating at 12 or 24 volts. In this case the armature shaft runs in ball-bearings at each end and has an inbuilt cooling fan at the commutator end, so that the cooling air flows first around the commutator and

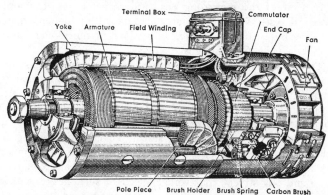

FIG. 39. The Bosch Type LJ/GTL ventilated, double ball-bearing type 8-inch diameter dynamo, rated at 12-volt, 500 watt and 24-volt, 1,000 watt

other parts which are subject to the greatest heating, e.g. the brushes. The air then passes through the apertures of the commutator end bearing and leaves again through the slots provided in the left end cover.

This type of dynamo, which is used on commercial vehicles, has a yoke or shell diameter of 8 in. and a terminal box, with protective cover, mounted on the yoke.

Dynamo Data. The automobile dynamo ranges in size from about 3-in. yoke diameter with an output of 70 to 90 watts for the smallest cars, to about 3·5 to 4·0 in. for the medium range of cars, with outputs of 200 to 270 watts, and up to 8-in. diameter for outputs of 1,000 to 1,700 watts, at 24 volts. Still larger dynamos are made for certain passenger vehicle purposes.

Four-pole dynamos, but with two brushes, are employed in the intermediate and larger sizes. In some instances four-pole, four-brush designs are used, to provide rather shorter commutators. In another example a six-pole dynamo with two to four brushes, of 8-in. diameter and 130 lb. weight, had an output of 1,600 watts, at 24 volts. The cutting-in speed was 1,100 r.p.m. and it gave 1,300 watts at 1,500 r.p.m.

In view of the increasing electrical demands on automobile electrical systems, there is a greater demand for higher dynamo outputs, but within the space limitations of the modern automobile or vehicle bonnet space.

To meet this demand, the direct-current dynamo has been developed to give greater outputs for a given size, so that for certain purposes, e.g. passenger vehicles with their big electrical loads, the limiting conditions in regard to size, weight and output appear to have been approached. One problem with such vehicles is that, due to their frequent stops, the engine is idling for relatively longer periods, so that the cut-out is out of action and no battery charging is possible. Further, in usually congested town and city streets, the average speed of such heavy vehicles —with their frequent stops—is of the order of 5 to 7 m.p.h. The dynamo speeds are often too low for adequate battery charge maintenance and there has been a tendency to increase the dynamo speeds, by using higher gear ratios, so that charging can be effected at vehicle speeds down to 7 or 8 m.p.h.

It is significant that over the period 1945 to 1960 the maximum operating speeds of British* automobile dynamos have increased from 7,000 r.p.m. to over 10,000 r.p.m., while the electrical load requirements at 12 volts have altered from about 12 amps. to 24 amps. for the standard equipment fitted to cars, and to over 30 amps. for equipment plus accessories.

When dynamo speeds are increased above the usual limits, commutation troubles may occur, e.g. sparking at the commutator due to the higher voltage. To overcome this trouble brushes of higher resistance have been used. In some instances commutating or interpoles are used, additionally, to give sparkless commutation with larger outputs over a wide speed range.

The tendency for commutator brush sparking at the higher speeds, due to the fact that this effect increases with the dynamo voltage, becomes important with 24-volt systems: in this connection electrographitic brushes or high resistance hard graphite brushes with brush pressures of 6 to 9 lb. per sq. in., giving a reduced voltage drop at the brush, are employed.

Dynamo Drives. The usual practice, on automobiles, is to employ a vee-belt and pulley drive from the engine crankshaft to the dynamo shaft. Usually, the same belt drives the cooling fan and water circulating impeller. Provision is made for taking up any excessive slack, due to belt stretch, by hinging the dynamo mounting so that it can be moved on a radius arm unit about a fixed hinge on the engine casing, in order to tighten the belt (Fig. 40). The belt should not be tensioned more than is absolutely necessary, or the dynamo bearings may wear appreciably in time. If too slack, there is a danger of belt slip so that the dynamo does not charge the battery satisfactorily. A common cause of failure to charge the battery is that due to the belt bottoming on the dynamo pulley. This fact is disclosed by the polished surface on the pulley groove, at its base.

As mentioned earlier, the modern dynamo drive pulley incorporates a cooling fan having radial blades which, when rotating

* Lucas.

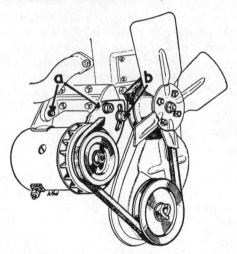

Fig. 40. Method of driving dynamo and adjusting belt tension (Renault). (a) Hinge bolts. (b) Slotted plate and locking nut

act as a suction fan to draw air through the yoke or casing from the commutator end, along the spaces between the armature of pole pieces and out through the fan blade openings.

Shunt Dynamo Characteristics. When the dynamo field is shunted across the (two) brush terminals, so as to place the field circuit in shunt or parallel with the armature circuit, then the effect of increasing the dynamo speed is to produce an induced electromotive force which increases with the dynamo speed. It can be shown that the voltage increases as the square of the dynamo speed (Fig. 41) provided the magnetic field is directly proportional to the exciting current, i.e. that the field resistance remains constant.

If, on the other hand, the exciting current is kept at a constant value, e.g. by a current supply from an outside source, then the effect of speed variation on the electromotive force is to increase this force in proportion to the speed, as indicated by the lower curve in Fig. 41.

It will be evident from these results that the shunt dynamo

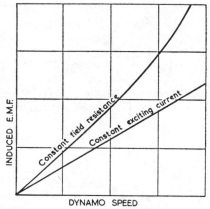

FIG. 41. Voltage characteristics of shunt-wound dynamo

characteristics shown are not suitable for a constant voltage supply system, so that some method of compensation must be provided in one of the dynamo circuits, in order to reduce the voltage tendency to increase with the engine speed, in an automobile.

It has been shown that the third-brush dynamo provides one solution, but has certain disadvantages, as mentioned earlier. It is proposed, now, to consider the other alternative method, which has more recently rendered the third-brush system obsolete, namely the *Constant and Compensated Voltage Control Systems.*

The Constant Voltage Control System. This method, which is still in use on a number of cars in the small, medium and larger classes, is based upon the principle of inserting, by automatic means, a resistance in series with the field winding, when the voltage reaches a certain value. In practice the operation is rather more complex, for reasons which will be explained.

The voltage regulator consists of an electromagnet, the coil of which is excited by armature current. At the upper end of the electromagnet's soft iron core a hinged spring-loaded armature member, shown as the vibrating bar, in Fig. 42, is mounted. A contact on the free end of the armature member can make

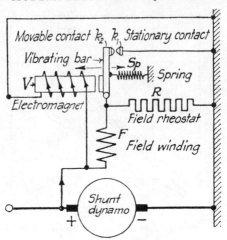

Fig. 42. Schematic arrangement of automatic single-contact voltage regulator

contact with a corresponding stationary contact, so that when the contacts are closed—due to the current increase in the electromagnetic coil—a resistance R is inserted in series with the field winding resistance F. When the dynamo is at rest or operating at its lower speeds the spring Sp closes the contacts k_1 and k_2, so that the upper end of the field winding is connected to earth, thus completing the field shunt circuit. Since the armature voltage is applied to the electromagnet winding, when the dynamo speed increases, this voltage will rise and eventually reach a predetermined value, such that the pull of the electromagnet overcomes the tension of the spring Sp and causes the contacts to open. The effect of this will be to insert the resistance R in series with the field resistance F, so that the total resistance becomes $R + F$. This increased field resistance will cause the armature voltage to drop and allow the spring to close the contacts, so that the voltage will increase again and break the contacts. This cycle of voltage changes causes the armature bar to vibrate several times a second; its effect is to maintain the voltage constant between two relatively small limits.

The voltage regulator shown schematically in Fig. 42 is known as the single-contact regulator and in a normal example will give rise to contact vibrations of 40 to 70 per second; in some designs a more rapid rate is provided. When the field winding has to be excited for longer periods in order to maintain the 'constant' voltage the ratio of the 'closed' to 'open' period is increased. When the speed increases, or the load is reduced, the contacts must remain closed for shorter periods so that the ratio of 'closed' to 'open' period is reduced. By careful design of this type of regulator it is possible to arrange for the duration period of the insertion of the field resistance R to adjust itself, automatically, in order to maintain the average voltage at a constant and predetermined value, independently of the load or dynamo speed.

Notes on the Constant Voltage System. In the relatively simple system shown in Fig. 42, the value of the inserted resistance R is made high enough to allow only a very small current to flow in the field winding. In this way the regulator contacts are protected against the inductive current surges which might otherwise occur should heavier current be allowed to flow in the field circuit.

With this type of regulator working in parallel with the automobile battery the voltage will be affected by the state of the battery. Thus, when the battery voltage is low the contacts remain closed for relatively long periods, so that a high charging current flows into the battery.

As the battery becomes charged, i.e. its voltage increases to the normal value, the regulator provides a smaller current which falls, eventually, to a mere trickle charge of 2 to 3 amps.

The principal disadvantage of this kind of voltage regulator is that for its satisfactory operation it requires a large output dynamo, so that should the battery be in the low charge state and a load be switched on, e.g. the full automobile lighting system, the voltage will fall still lower. In order, therefore, for the regulator to maintain its 'set' voltage, a very heavy current will flow through the armature and battery, so that there would be a grave risk of burning the armature coils. It is for this reason that the modified type of regulator, known as the *Compensated Voltage Control*, was developed by Messrs. Lucas Ltd, but before describing this regulator, reference will be made to an interesting modification of the single contact regulator, described earlier, namely, the *Double Contact Regulator*.

The Double-Contact Regulator. Instead of using a single set of contacts, as in Fig 42, a pair of such contacts can be employed with advantage in certain applications. Thus, when a dynamo has to operate at high speeds, such that the inserted resistance is insufficient to reduce the exciting current value, it then becomes necessary to cut out the field winding, intermittently. This is effected by means of a second pair of contacts, k_2 and k_3 (Fig. 43), which come into operation when the electromagnetic pull is sufficiently great. It will be seen that when these contacts close, the field winding F is short-circuited.

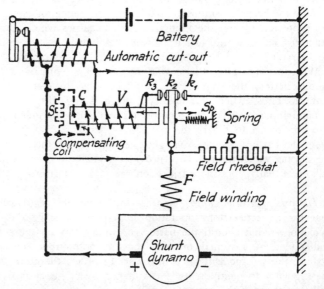

FIG. 43. The double-contact voltage regulator, with compensating winding C and adjustable shunt resistance S

THE CHARGING SYSTEM

As before, the contacts k_1 and k_2 operate at the lower-to-normal speeds to insert the resistance R but when the dynamo speed increases appreciably, or the load becomes very small, the contacts k_1 and k_2 will remain apart permanently, while the other pair k_2 and k_3 vibrate, putting the field resistance R in and out of circuit. When these contacts touch the resistance R is connected in parallel with the electrical load; when not in contact, R is in series with the field winding F.

With the double-contact regulator the regulating resistance can be of lower value than for the single-contact type, with the result that the contacts have a longer life, while the exciting currents can be of higher value. The double-contact type will control the output voltage even at high speeds.

The double-contact regulator, once standard equipment on various makes of cars and commercial vehicles is more complicated and needs very careful adjustment and has since been superseded by the single-contact unit, except in the case of commercial vehicles. Reference to a typical compensated voltage control double-contact regulator is made, later, in this chapter.

The Compensated Voltage Control Regulator. Before describing this popular regulator its advantages might first be emphasized:

The regulator overcomes the previously mentioned drawbacks of the constant voltage type by providing for automatic variation in the actual or operational voltage setting so that excessive currents, due to the difference between the dynamo terminal voltage and the back E.M.F. in the charging circuit cannot occur.

The regulator, in this case has an additional winding, which is responsive to current; thus, the winding is in series with the armature coils and takes the main charging current. The previous voltage winding is retained and the current winding is arranged so that it assists the voltage winding in separating the moving and fixed contacts of the regulator. When the cell voltage is low, e.g. 1·8 volts, it follows that with the full electrical load in circuit a very high current would be needed, as explained earlier. To prevent this it is the modern practice to arrange for the regulator to provide what is known as a *drooping characteristic curve*, such

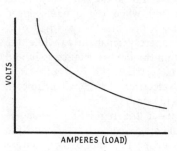

Fig. 44. Drooping voltage characteristic of regulator

that the voltage falls as the load increases (Fig. 44). This gives a tapering charge for the battery, the current falling off during the charging period. Another modification is incorporated in the automatic voltage control regulator, to enable the load imposed by the vehicle's lamps to be carried by the dynamo so that the battery can be kept fully charged. This is effected by the addition of one or two further turns on the compensating winding so that the lamps and other equipment electrical load current can be taken through this additional winding. In order to simplify matters it is usual to make both the windings in one and to tap off the connection to the main charging circuit at the correct place.

The result of this modification is that the generator cannot be overloaded when the full electrical load is switched on, with a 'low' battery; the battery will also remain fully charged and 'float' in the charging system when the full load is 'on'.

The regulator is usually adjusted to give from 15 to 16 volts at no-load; this is the voltage of the fully-charged battery. When charging a 'low' state battery the voltage drops to about 13·5.

It will be seen that when there is an electrical load in the circuit, with current in the series coil, the voltage will fall as the load rises.

Typical Compensated Voltage Regulator Circuit. The arrangement of a regulator which embodies the features mentioned in the previous paragraph is shown, diagrammatically, in Fig. 45. Since the modern regulators also contain the cut-out unit on their mountings, the wiring of the cut-out has been included; it is shown on the left hand side of the diagram. The electromagnet core, shown in the central part of this diagram, is

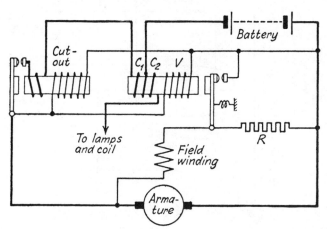

Fig. 45. Single-contact automatic voltage control circuit, including cut-out

provided with the usual armature brush connection voltage coil, i.e. the voltage of the main charging current, at V and the series coil at C_1, C_2. One portion C_1 of this coil is placed in the external circuit of the dynamo while the other, C_2 is in the lead from the battery to the electrical components of the car. As explained, earlier, the combined effects of the three coils govern the movement of the electromagnet's hinged armature with its contact to produce the desired regulator action under different load, speed and battery conditions.

Summarizing the operations of this regulator, it can be stated that:

(1) With a discharged battery and no electrical loads in circuit the dynamo will develop its full output, due to the coil C_1.
(2) By arranging for the main electrical load current, e.g. the lamps, to go through the coil C_2 the voltage is reduced so as to prevent the generator output from increasing. The generator can thus be maintained at full output, independently of the electrical load.
(3) Both overcharging and overloading are insured against.

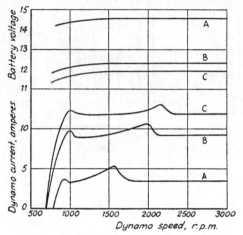

Fig. 46. Characteristic regulation curves for voltage control unit

Characteristic Curves. The performance of a compensated voltage control system in an automobile battery charging circuit is shown by the three pairs of voltage and current graphs given in Fig. 46. The curves marked A, B and C correspond to an almost fully charged, almost discharged and a fully discharged battery, respectively.

It will be seen that when almost fully charged, the dynamo supplies only a small charge, of 3 to 5 amps. When nearly discharged the dynamo charging rate increases to about 8 to 11 amps. and when fully discharged to 11 to $12\frac{1}{2}$ amps. The charging rate, therefore, automatically adjusts itself to the state of the battery charge.

Comparison with Third Brush System. The marked advantages of the compensated voltage control method over the third brush method is shown by the characteristic curves in Fig. 47. in which 'constant' here refers to 'compensated'.

It will be seen that with the third brush system the charging current is low when the battery voltage is low, i.e. the battery is in a 'low' charge state. As the battery voltage increases the charging rate also increases, thus tending to overcharge the battery.

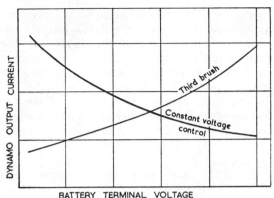

FIG. 47. Comparison of two and three brush charging systems

With the compensated voltage regulator, when the battery voltage is low the dynamo output, or charging rate is high. As the battery voltage rises the charging rate is reduced progressively, as indicated by the drooping characteristic curve.

Effect of Operating Temperature. The operation of the dynamo and regulator are affected by their actual temperatures. Thus, the dynamo when cold can produce a greater charging current than when hot. This is an advantage, in automobiles, since the cold dynamo can provide the greater output need for charging a cold battery—for the resistance of the battery acid is greater when cold. Further, as the starting motor requires a considerable current, the dynamo is able to provide this current better, when it is cold.

In regard to the regulator the resistance of the regulating voltage coil is lower when it is cold than when at normal temperatures. As the coil temperature rises the resistance also increases; this causes the regulator to control at a higher voltage. In this connection, the resistance of a copper conductor increases by 0·4 per cent for every °C above 20° C, so that if the coil temperature rises to 50° C, the resistance will increase by 0·4 × 30 = 12 per cent. If the exciting voltage is constant, this temperature rise will result in a reduction in the exciting current of 12 per cent.

With a lower exciting current the magnetic pull on the vibrating armature is reduced so that a higher voltage is needed to open the contacts and the battery in consequence receives a greater charging current when hot than if cold. The change from winter to summer conditions therefore necessitates some method of temperature compensation on all accounts.

The simplest compensating device is that of the bimetallic strip, consisting of two strips of different metals, welded together. If the strip is flat when cold it will bend, progressively, as the temperature rises. This relative movement is utilized in temperature compensating devices.

In the case of the compensated voltage regulator the spring tension of the hinged armature member is made variable, automatically, by means of a bimetallic device, so that the controlling voltage is higher when the strip is cold than when hot, thus providing for a greater dynamo output when cold.

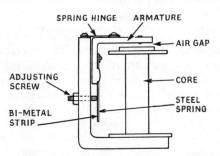

Fig. 48. Bimetallic strip temperature compensator for voltage regulator

Fig. 48 illustrates the application of a bimetallic strip to a voltage regulator. The normal steel spring, in strip form, is shown on top, with the armature plate on its left hand side. The bimetallic strip is attached to the bent part of the armature plate and, at its lower end there is an adjusting screw for the initial setting of the regulator voltage according to the temperature. The force on the adjusting screw is reduced as the temperature rises, so that the voltage setting of the regulator is reduced.

The cut-out coil is also affected by temperature changes, since the resistance increases as the temperature rises, thus reducing the magnetic pull on the armature. This effect can be compensated against by using an armature bimetallic spring, so that the operating voltage of the cut-out is maintained constant.

Typical Temperature-Voltage Values. In the case of a typical 12-volt British voltage regulator, the following readings were obtained during tests:

Air Temperature °F	50	68	86	104
,, ,, °C	10	20	30	40
Voltage (Open Circuit)	14·5–15·1	14·2–14·8	13·9–14·5	13·6–14·2

These voltages were used later to set other regulators manufactured for the same dynamo. In this connection the manufacturers of electrical equipment always provide the open-circuit voltage settings for different operating temperatures. A suitable screw adjuster is located on the regulator for this purpose.

The standard temperature-voltage specification for the American Delco-Remy compensated voltage regulator is as follows:

Air Temperature °F	45	65	85	105
,, ,, °C	7·2	18·3	29·4	40·6
Normal Voltage	14·5–15·6	14·4–15·4	14·2–15·2	14·0–14·9
Air Temperature °F	125*	145	165	
,, ,, °C	51·7	62·8	73·9	
Normal Voltage	13·8–14·7*	13·5–14·3	13·1–13·9	

The air temperature is taken at about ¼ in. from the regulator cover. It is important that the regulator is brought to its proper operating temperature before taking the temperature measurement; this usually necessitates operating the regulator for at least 15 minutes, with its cover in place.

Position and Cooling of Dynamo. The position of the dynamo on an automobile should be chosen carefully, so as to fulfil as far as possible the following requirements:

(1) The commutator and brush gear should be accessible, for servicing purposes.
(2) The dynamo should be kept as cool as possible.

* Published specifications.

(3) The bolts and nuts provided for belt tensioning purposes, normally hold the dynamo mounting and so should be readily accessible.

In regard to the important subject of cooling the dynamo—since, as stated earlier, the dynamo output falls off with increase in its temperature—the dynamo should be placed as nearly as possible in the radiator fan cooling air stream; in this connection the modern ventilated dynamo should provide for good access conditions for the ventilator air.

The results of some tests[*] on a 12-volt dynamo mounted in a position where there was no cooling wind and again, on a cylinder block just behind the radiator fan showed that in the first case, the maximum cold output was 11·5 amps. at 13 volts; but when it became heated up, the output fell to 8·6 amps. at 13 volts. In this case the outside yoke and commutator temperatures were 93° C and 147° C, respectively.

In the second case, the cold output was the same, but in the heated condition, when under the influence of the cooling air stream, the output was 10·2 amps. at 14·5 volts; in this case the yoke and commutator temperatures were only 32° C and 78° C, respectively.

Typical Compensated Voltage Control Regulator. Fig. 49 illustrates a widely-used regulator assembly, namely, the Lucas Type RB.106, in which both the voltage regulator and cut-out units are mounted on the same base, the cut-out being on the right hand side. In use, a cover encloses the whole assembly, except for the terminal connections. The voltage regulator unit belongs to the split-series winding class, described earlier, the electrical diagram of which is shown in Fig. 50. It will be observed that the terminal connections at the lower parts of the two diagrams are lettered similarly. In regard to the split-series winding of the regulator coil unit, the terminal A is connected to the battery and the terminal $A1$ to the lighting and ignition switch.

The regulator armature embodies the temperature compensation device described previously; this is located above the coil

[*] E. A. Watson.

THE CHARGING SYSTEM 95

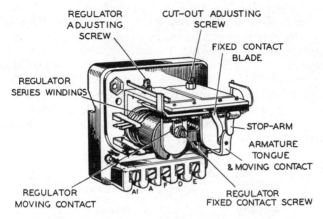

Fig. 49. Lucas RB.106 cut-out and regulator assembly, with cover removed

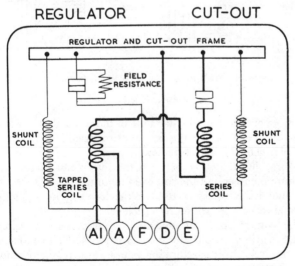

Fig. 50. Circuit diagram of RB.106 cut-out and regulator

and its regulator adjusting screw—which corresponds to that shown in Fig. 48—is clearly indicated. Similarly, there is an adjusting screw for altering the mechanical setting of the cut-out contacts gap; in the present example the correct gap, when the armature is in the free position is ·018 in. while the gap between the armature or angular hinged member and the fixed stop arm above is between ·025 and ·030 in.; this corresponds to the closed contacts position of the armature. Any alteration of this gap can only be done by bending the stop arm. The gap clearances mentioned are shown in Fig. 51.

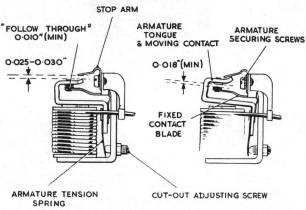

FIG. 51. Method of adjusting cut-out

Fig. 52 shows the regulator unit in more detail and also illustrates the methods of adjusting the two essential gaps, namely, that between the coil core end face and the moving angled armature, and the other between the lower end of the vertical armature member and the coil fixed mounting.

Should it be necessary to alter the former gap, it should be adjusted to ·015 in. with a feeler gauge.

The voltage regulator adjusting screw is used to alter the voltage setting, should this not agree with the manufacturer's temperature-voltage specification values. If the regulator adjusting screw is turned clockwise the voltage reading is increased and if anti-clockwise, decreased.

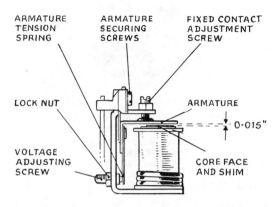

Fig. 52. Method of adjusting voltage regulator

For this test, terminals A and $A1$ (Fig. 50) are joined together and the negative and positive leads of a 0–20 volt moving coil voltmeter are connected to the D and E (earth) terminals, respectively. The voltage at which the reading becomes steady, when the engine is accelerated, slowly, from idling speed, is the correct one.

Current and Voltage Regulator. In this type of regulator, in addition to the voltage coil of the compensated voltage regulator there is an additional coil in series with the dynamo output, so that it takes the full current. This coil is provided with a vibrating armature and contacts, similarly to the voltage coil. The nett effect of using both a voltage and a current controlled system is that both the voltage and current values are controlled to suit the electrical loading and battery condition. Thus, with current and voltage control (C. and V.C.) the current control arranges for the battery to be charged up to the predetermined voltage; then, the voltage control comes into action and, eventually, the current decreases to the trickle charge value.

In practice the three units, namely, the cut-out, voltage control and current control members are mounted on a common case, and provided with a dust cover over all the units.

Fig. 53 shows the Lucas RB.310 current and voltage control regulator, with the cover removed. The terminals marked B, F

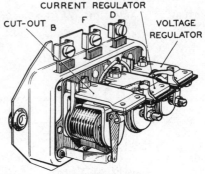

FIG. 53. Lucas RB.310 current and voltage control and cut-out unit

and D refer to the corresponding cable connections to the battery, field and dynamo output. In this case the mounting plate is vertical, the coils being horizontal. In the Auto-Lite, Delco-Remy and Bosch regulators the coils are mounted vertically.

Fig. 54 shows the circuit arrangement of a current and voltage control unit, together with that of the cut-out unit, since it is mounted on the same base and has certain common leads. It should here, again, be emphasized that in all such regulators, the cut-out is an independent electrical unit and does not, therefore, affect the action of the regulator.

Referring to Fig. 54, the current unit is wound with a few turns of a heavy-gauge wire, since it takes the full current output of the dynamo. As before, the voltage unit is wound with a much larger number of fine wire turns, since its current value is relatively small. Two resistances R are provided, one for the voltage and the other for the current coil contacts system. The operation is as follows:

When the dynamo speed is increased from the idling condition, the cut-out contacts first close, to allow the dynamo output current to flow through the cut-out's closed contacts and also the coil of the current control unit. When the current reaches its predetermined value, its contacts separate and in doing so insert the resistance R on the left hand side, in series with the dynamo field, thereby reducing the current output and, therefore, the pull on the current unit's armature so that the spring closes the

THE CHARGING SYSTEM 99

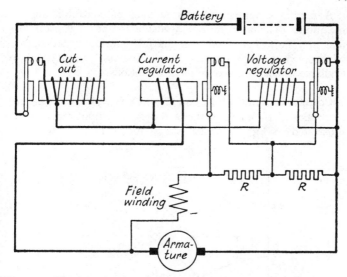

Fig. 54. Circuit arrangement of typical current and voltage control system

contacts and the current from the dynamo again increases. This cycle of operations is repeated at a relatively high frequency, namely between 50 and 200 times per second. Thus, the current regulator behaves, mechanically, in a similar manner to the voltage regulator in its vibrating armature method of control, but in this instance, of the current strength.

The voltage regulator coil takes the full voltage of the dynamo current and its purpose is to insert the right hand resistance R whenever the voltage reaches its predetermined maximum value. This results in a reduction in the current output of the dynamo. With the maximum voltage of the dynamo applied to a 'low' battery, the charging rate due to the current regulator would be too great over the required battery charging period. The effect of the voltage coil, however, is to reduce the charging current, so that as the charging operation proceeds the current is reduced in value, finally, dropping to the trickle charge of 3 to 4 amps.

A further advantage of this method of control is that the maximum dynamo output is available for any electrical loads that may be switched on, e.g. the full lighting system, heater, windscreen wiper, radio, etc. Thus, the full output is always available under road requirement conditions.

Comparison of Co.V.C. and C. and V.C. Systems. The characteristics of these two methods of dynamo output regulation are illustrated in Fig. 55 which shows the manner in which the

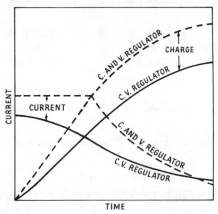

FIG. 55. Performance graphs for Co.V.C. and C. and V.C. charging systems

charging currents of the two methods of regulation vary. It will be seen that with current and voltage regulation the current-limiting relay enables the battery to be charged at the given output current up to a predetermined voltage, after which the voltage regulator comes into action and the current fall to a trickle-charge occurs. This constant rate of charging is maintained until the battery is almost charged; further, the maximum dynamo output is available, whatever the car's electrical demands, or load. With this system, control with the current regulator is so precise that the dynamo output cannot increase, even if a cable is short-circuited, so that it is *unnecessary to fit a fuse* in the generator circuit.

The higher charge rate of the current and voltage system is shown by the upper curves in Fig. 55 and the current maintenance at a given value, is indicated by the dotted line, parallel to the time base. With the constant voltage system the charging current of the regulator falls off continuously down to the final trickle charge value. This greater ability of the current and voltage system to balance heavy electrical loadings, which is more marked in the larger sizes of dynamos and higher loads, renders the system particularly applicable to commercial and passenger vehicles, as well to the larger cars, e.g. the American ones—in which this system is standard.

Temperature Compensation. The voltage coil armature spring is provided with the bimetallic strip method of compensation, for temperature changes, similarly, to that of the Co.V.C. unit. The actual voltage-temperature values, however, are usually different from those of the latter unit.

Thus, in the example of the American Delco-Remy C. and V.C. regulator the regulator voltages tend to be rather higher at the higher temperature values than for the corresponding Co.V.C. regulator.*

Regulator Contacts. In order to reduce the wear of the vibrating contacts to a minimum tungsten is used for these contacts in single-contact type regulators and silver or tungsten-silver in double-contact type regulators.

In cases where large dynamos are employed—as on commercial and passenger vehicles—the field coils are sometimes divided so as to give two parallel circuits, each of which has its own regulator, so as to reduce the wear on the regulator contacts. In this connection, four-pole dynamos experience heavier loads than two-pole ones.

Induced Voltage Surges. When the current and voltage regulators are operating there is, in each case, a sudden reduction in the dynamo current, due to the insertion of a resistance in series with the field winding. This current reduction causes a

* Vide page 93.

surge of induced voltage in the field coils, due to the change in strength of the magnetic field. The resistances in the regulator circuits provide for a partial dissipation of the surge effects and therefore reduce any tendency to excessive sparking across the contacts of the armature units.

The Bucking or Frequency Coil. This coil is employed in connection with the operation of the voltage regulator contacts, to give a faster breaking action and increase the frequency of the vibration, thus providing a smoother dynamo output control.

The bucking coil is wound in series with the two sets of contacts in the dynamo field circuit. The winding, consisting of a few turns of thick copper wire, is arranged to assist the shunt coil of the voltage regulator, when the output current flows around it.

Using a Single Field Resistance. Instead of employing the two resistances, marked R in Fig. 54, a single resistance can be employed, since the voltage and current regulators, except for the short change-over period when both regulators may be in action, normally operate singly. Fig. 56 shows the Lucas current and voltage circuit, for the regulator illustrated in Fig. 53. The single resistor is shown connected between F and the voltage coil armature, while the frequency, or bucking coil is also shown.

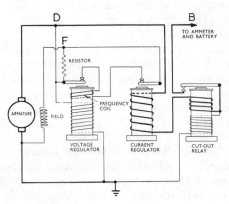

FIG. 56. Circuit diagram for Lucas current and voltage control system

THE CHARGING SYSTEM

Single Core Regulator and Cut-out. Instead of using two separate iron cores for the compensated voltage control and cut-out units it is possible to employ one core only, as depicted in Fig. 57. This arrangement is employed in certain of the Bosch regulators and results in a more compact combination.

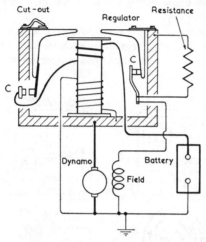

FIG. 57. Single core cut-out and voltage control unit

Referring to Fig. 57 the core is wound with a series and also a voltage winding with the voltage armature contacts on the right hand side, and the cut-out controls on the left. The magnetization of the core is initiated by the shunt coil and when of sufficient strength causes the voltage and cut-out armatures to be attracted to the core, thus connecting the cut-out contacts, to close the dynamo charging circuit and opening the regulator contacts, in a similar manner to that described earlier. The combined unit is designed so that the cut-out armature is attracted when the open circuit voltage is 13, while the voltage regulator opens at 15 volts, i.e. after the cut-out contacts have closed. The series coil shown not only provides for voltage compensation but also provides a strong magnetic field, to hold the cut-out contacts firmly together.

It is possible to simplify, further, this type of single core unit by using a single armature for both the cut-out and voltage regulator units, as is done in the case of the Bosch combined unit shown in Fig. 60.

Delco-Remy C. and V.C. Regulators. These regulators are widely used on many makes of American cars and to a certain extent in this country.

Fig. 58 illustrates a typical three-unit regulator having the cut-out, current and voltage regulators mounted on the same base and fixed to the back of the dash by a rubber mounting, to damp out vibration effects.

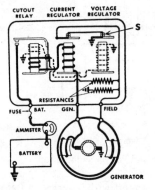

FIG. 58. Delco-Remy single contact C. and V.C. regulator

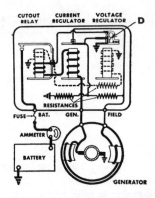

FIG. 59. Delco-Remy double contact C. and V.C. regulator

The regulator is made in two principal designs, namely, for the single and double contact circuits as illustrated in Figs. 58 and 59, respectively. In general, the single-contact regulator is made for dynamos with maximum outputs of 25 to 35 amps., while the heavier duty model is used for dynamos with maximum outputs of 40 amps. and upwards. In a typical example, namely, the more recent Buick 250 to 325 b.h.p., vee-8 engine cars, the 70 amp. hr. battery is used in conjunction with a 35 amp. dynamo, single-contact C. and V.C. regulator, while in air-conditioned models a 45 amp. dynamo and double-contact breaker regulator is used. The dynamos are of the ventilated type and the armature shafts

THE CHARGING SYSTEM

are mounted on ball-bearings at each end. The dynamo has a double vee-type pulley, and uses two vee-belts to drive the armature shaft, from the crankshaft pulley.

The cut-out unit follows the usual practice in having both a shunt and series winding on the same core.

The current regulator (Fig. 58) has one series winding of heavy gauge wire, which is connected to the series winding in the cut-out unit. Above the winding core is an armature, with a pair of contacts held together by spring tension when the current regulator is not operating. When the dynamo current increases to the predetermined maximum, the armature is pulled towards the core, thereby putting the upper resistance, shown in Fig. 58, in series with the dynamo field circuit. As explained earlier, the current falls in value and the contacts close again. This cycle of operations is repeated at a high rate, giving rise to the armature vibration control which regulates the current output.

The voltage regulator has a similar type of armature with a single pair of contacts S which are closed when the voltage across the brushes is low and opened when the voltage reaches the given maximum setting. It acts on the vibrating armature principle to govern the voltage according to the electrical load and battery requirements. The voltage regulator has a bimetallic armature hinge for thermostatic temperature correction purposes.

The current and voltage regulators do not act at the same time, for when the current draw is large the dynamo voltage is too low for the operation of the voltage regulator, so that the current regulator limits the dynamo output; when, however, the current demands are small the voltage of the dynamo is allowed to increase to the voltage regulator's operating value. This reduces the dynamo output leaving all control dependent upon the voltage regulator.

In the case of the higher output dynamo with the double-contact (D) voltage armature unit a third resistance is connected between the cut-out base plate and the voltage regulator winding, to reduce the magnetic pull of the voltage winding; this resistance is indicated by the broken line above the full live resistance on the right hand side, in Fig. 59. As mentioned earlier, these resistances in addition to their dynamo output control functions, also tend to reduce electrical surges in the field coils, when the contacts vibrate to insert or take out the additional field resistances.

The Bosch Regulator Units. Various designs of voltage regulators with inbuilt cut-out units are made by the Bosch Company; these are available in both the single- and double-contact models.

A typical compensated voltage regulator and cut-out unit, suitable for incorporating in or mounting on the dynamo is illustrated in Fig. 60. The special feature of this regulator is that

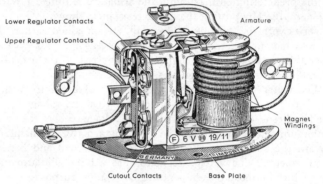

FIG. 60. Bosch compensated voltage regulator and cut-out. Type F

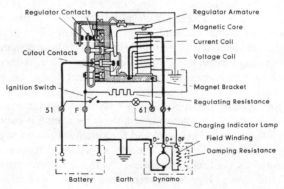

FIG. 61. Circuit diagram of Type F regulator and cut-out unit

THE CHARGING SYSTEM 107

it employs the same single iron core for both the voltage regulator and the cut-out with a combined armature member.

A complete circuit diagram for this regulator unit is shown in Fig. 61. The principle of operation of both the cut-out and voltage regulator is the same as for the separate core regulator and can be followed from the circuit diagram and the information given earlier on the single core regulator. It may here be mentioned that the voltage control gives a drooping characteristic voltage curve, which is particularly advantageous for automobile charging circuits. A typical three-element current and voltage control, with cut-out unit, as used on certain American and some Continental cars, is shown in Fig. 62, while the corresponding circuit diagram is given in Fig. 63.

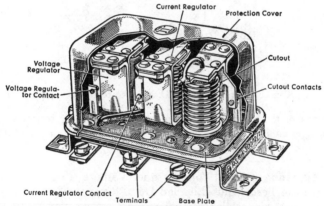

FIG. 62. Bosch C. and V.C. regulator and cut-out unit

The voltage control unit is of the double-contact kind, while the current control member has a single contact for its armature. In the conventional circuit the field winding of the dynamo is connected to an insulated carbon brush on the dynamo while the negative pole is earthed.

The principle of operation of this regulator is the same as that previously discussed, namely, by the insertion and withdrawal of resistances in the field circuit of the dynamo by the vibratory

108 MODERN ELECTRICAL EQUIPMENT

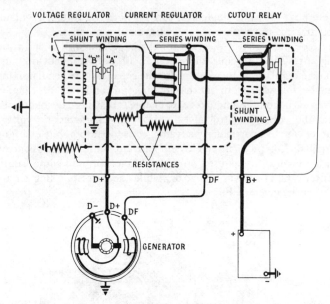

Fig. 63. Circuit diagram of Bosch C. and V.C. regulator and cut-out unit

action of the armatures. A common compensating resistance is incorporated for both the voltage regulator winding and the cut-out voltage winding.

Adjusting Cut-in Voltage and Voltage Limit. In the American Bosch, Auto-Lite and Delco-Remy three-unit C. and V.C. regulators, while screw adjustments are sometimes provided for altering the contact points gaps, it is standard practice to adjust the cut-in voltage by bending the armature stop, and the voltage and current regulator contact opening voltage and current values by bending the bottom spring anchoring arm, or hanger upwards or downwards with a special tool made for this purpose.

Fig. 64 illustrates the combined regulator and cut-out unit, with cover removed, of the American Bosch three-unit regulator, as used on the Ford Company's cars. The springs that hold the contacts open, when the engine is idling, are connected at their

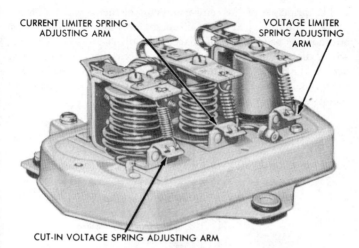

Fig. 64. Heavy duty C. and V.C. regulator showing adjustment items

lower ends to brackets, or anchoring arms attached to the base member.

To alter the cutting-in voltage of the cut-out, the arm is bent upwards to increase the voltage and downwards to decrease the voltage.

To adjust the voltage regulator voltage limit the anchoring arm is bent upwards to increase the voltage and downwards to reduce the voltage; the same method applies to the Auto-Lite regulator.

With both the Bosch and Auto-Lite heavy duty type regulators the current limit on the current regulator can be increased by bending the anchoring arm downwards and reduced by bending the arm upwards, using the special bending tool supplied.

Semi-conductor Type Regulator. A more recent automobile type regulator employs a semi-conductor element, developed by Bosch, with what is known as a *pn*-junction. The regulator has a similar characteristic curve to that of the current and voltage regulator, but dispenses with the usual current control member.

The principle of the semi-conductor element, which consists of germanium doped with antimony or indium is that when antimony is used it produces an excess of negative charges and, when the indium alloy is used an excess of positive charge occurs. The junction of the negative or *n*-type and the positive, or *p*-type material is the essential element of this regulator member.

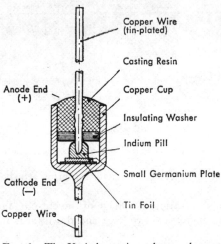

FIG. 65. The Variode semi-conductor element

Fig. 65 illustrates the arrangement of the various components of the Variode unit, based upon this *pn*-junction principle. When subjected to a low voltage current in the 'forward' direction the unit allows only a weak current to flow through, but as the voltage is increased then the current is increased at a much more rapid rate, as shown in the characteristic curve in Fig. 66. In this way, the overall control curve is similar to that of the C. and V.C. regulators.

The Variode regulator, in appearance, resembles the two-unit cut-out voltage regulator, but has no current control unit. Instead, the current control winding is replaced by a control winding which at one side is connected with the copper cup cathode and at the other side through the cut-out voltage winding and armature,

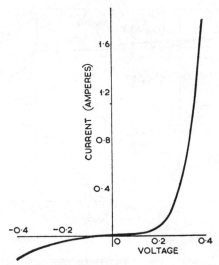

Fig. 66. Characteristic curve of Variode unit

as shown in Fig. 68. The Variode is located on the base of the regulator and cut-out units, and occupies very little space.

Operation of pn-Regulator. The Variode regulator operates in a similar way to the compensated voltage control regulator, giving the same type of drooping voltage characteristic curve, as also with the C. and V.C. regulator. It is mainly in the variation of the voltage to suit the dynamo load that the Variode regulator differs from the other two types of regulator. The differences between the three regulator characteristic curves are shown in Fig. 67, for a 6-volt system.

Referring to Fig. 68 a weaker conductor is connected in parallel with the main current conductor, namely, from $D+$ through the cut-out current winding to the cut-out contact, the former leading to the Variode element to the control winding on the regulator element and receiving the voltage drop that occurs in the main current conductor, due to the resistance.

At low dynamo loads the voltage drop is very low, so that only a very weak current goes through the Variode. At a predetermined

112 MODERN ELECTRICAL EQUIPMENT

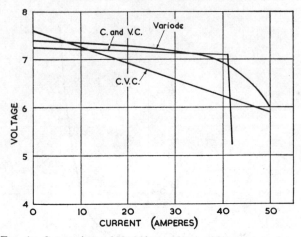

Fig. 67. Comparison of Variode regulator with other regulators

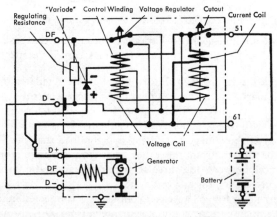

Fig. 68. Wiring diagram of Variode regulator and cut-out

voltage drop, corresponding to a given dynamo load the current in the control winding rises considerably. The resistance in the main current conductor is selected so that the Variode acts fully at the maximum permissible current value and the magnetic field generated by the control winding causes a rapid decrease in the dynamo voltage thus protecting the dynamo against overloading; it will be understood that this is a similar result to that obtained by the current coil of the C. and V.C. regulator.

Temperature Effects. As in the case of the ordinary voltage regulator, the dynamo voltage increases with falling temperature and falls with increasing temperature, so that the voltage member is provided with the usual bimetallic temperature compensation device. An advantage of the Variode is that the voltage is reduced on the cold regulator only at a current intensity which is considerably above the peak value allowable for the dynamo; this does not, however, affect the cold dynamo adversely. If the temperature of the dynamo and regulator rises, the current is limited to the allowable value. Thus, the incorporation of the Variode in the regulator enables the dynamo to be better utilized under heavy load conditions, and under city driving purposes, with frequent stops and low average speeds allows the battery to be kept in the well charged state.

Typical Electrical Specifications. The following electrical data refers to Lucas equipment fitted to cars with engines of $1\frac{1}{2}$ litre capacity, and for the compensated voltage method of regulation:

Battery. 38 ampere-hour at 10 hour rate and 43 ampere-hour at 20 hour rate.

Dynamo. Two-brush, two-pole shunt wound ventilated type. Field resistance, 6·0 ohms. Maximum output 19 amperes at 1,900–2,100 r.p.m. at 13·5 dynamo volts. Brush spring tension, 22–26 oz. Rotation, clockwise from the drive end.

Cut-out and Regulator. Cut-out contacts close at 12·7–13·3 volts, at 1,300 r.p.m. (dynamo). Drop-off (contact opening) voltage 8·5–11·0 volts. Reverse current, 3·0–5·0 amperes. Regulator open circuit settings at 3,000 r.p.m. with regulator at following ambient temperatures:

10° C (50° F)—16·1–16·6 volt
20° C (68° F)—16·0–16·5 ,,
30° C (86° F)—15·9–16·4 ,,
40° C (104° F)—15·8–16·3 ,,

The following information refers to a typical American electrical installation in a car having full electrical equipment, including power-operated seats and windows, and refrigerating-heating air conditioning. The eight-cylinder engine is of the 250–300 b.h.p. type and the car has automatic (Hydra-Matic) transmission. The current and voltage control system is employed.

Battery. 60 ampere-hour, at 20 hour rating. 9 plates per cell.

Dynamo. Delco-Remy, two-brush, two-pole shunt wound. Cold output, 35 amperes at 14 volts at 2,500 r.p.m. Field current draw, 1·7–1·8 ampere at 12 volts (80° F.).

Cut-out. Contacts close at 11·8–13·5 volts. Cut-out relay air gap, ·020 in. Cut-out point opening, ·020 in.

Regulator. Voltage regulator normal range, 13·8–14·8 volts. Voltage regulator air gap ·075 in. Current regulator setting 32–37 amperes. Current regulator air gap, ·074 in.

Alternating Current Generators. With the ever-growing tendency to increase the electrical equipment, and therefore the electrical loading of automobiles and also commercial vehicles, the demands on the direct current dynamo are such that only by increasing the size and weight and also running the dynamo at higher speeds, can the necessary output be obtained.

As mentioned earlier there is a definite limit to the maximum speeds of the dynamo due to brush and commutation limitations, so that it has become necessary in certain cases to employ the alternating current dynamo or generator, with its known advantages but, usually, greater initial outlay. Before describing typical automobile A.C. generators, it may be of interest to outline their advantages.

In the first place, for a given output, i.e. wattage, the A.C. generator can be made *appreciably lighter and smaller*. Thus, in a typical case a conventional D.C. dynamo of nearly 8 in. length and 5 in. diameter, had a maximum output of 35 amps. at 13·5 volts, with a maximum possible operating speed of 9,000 r.p.m. and weight of 26½ lb. The alternative A.C. generator, with its current rectifier unit, had an output of 45 amps. at 13·5 volts (D.C.) and a maximum speed of 11,000 r.p.m., for a total weight of just under 18 lb. It was just under 6 in. in length, but of 6 in. diameter.

The *maintenance attention is much reduced*, owing to the use of light slip ring brushes, there being no commutator and heavy brush gear. The life of the brushes before attention is required is many times that of the commutator-type brush. Since, also, there is no commutator with its temperature and wear problems, this is an additional advantage.

A *high output can be obtained at low engine speed*, so that when the engine is idling the battery can be charged; this is not possible with the ordinary dynamo with its cut-out. For larger passenger automobiles and buses, with their relatively low average city speeds and frequent stops, the A.C. generator has very definite advantages.

Silicon Rectifiers. The development of automobile generators of high output from small units has been possible largely on account of the silicon rectifiers now available. These rectifiers, themselves of small dimensions enabling them to be mounted in the end cover of the generator, have a high resistance to current flow in one direction and a low resistance to flow in the other direction. When the correct current polarity is used, the rectifier allows relatively large currents to flow from the generator to the battery, in the low resistance direction, but on account of the high resistance to flow in the other direction, prevents current flow from the battery to the generator. The alternating current is transformed into direct current by the rectifier.

An important advantage of this rectifier is that since current flow is prevented from the battery to the generator, *no cut-out unit is required*.

Applications. The A.C. generator has been in use for some appreciable time for bicycle and motor-cycle lighting purposes. More recently, it has been used as standard equipment on the Chrysler cars, and is available commercially both for cars and heavier vehicles, by Lucas, Delco-Remy, C.A.V. and some other manufacturers.

Examples of A.C. Generators. The Lucas 2AC alternator is available for vehicles which require greater electrical demands than are provided for by the standard dynamos, e.g. public service vehicles and automobiles with full electrical equipment, including two-way radio equipment. It is particularly suited to the requirements of vehicles that have to make frequent use of the motor starter and those operating at low average speeds.

The Lucas generator (Fig. 69) incorporates a stator output winding and a rotor field winding energized through a pair of slip rings. In view of the absence of commutator and its brush

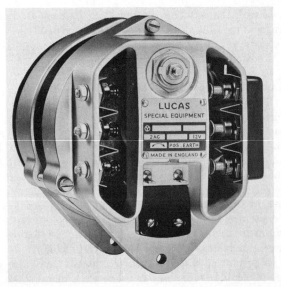

FIG. 69. Lucas Model 2 AC alternating current generator

gear higher driving pulley ratios are employed; so that the output is appreciably greater at lower engine speeds.

An important feature of this generator is the incorporation of the current rectifier in the fixed end casing member. This takes

FIG. 70. Method of driving A.C. generator

advantage of the more recent *silicon diode rectifier*, with its relatively high current capacity and its small dimensions.

Six of these small rectifiers are built into the slip ring end bracket, i.e. at the slip ring end where they are cooled by an air stream, induced by a 6-in. diameter cooling fan in the driving pulley. The end bracket acts also as a heat sink, while additional cooling surfaces are provided by thin copper strips fitted beneath the rectifiers.

The *generator stator*, of 6-in. outside diameter, comprises a 36-slot, 3-phase delta-connected winding on a ring-shaped lamination pack between the two cast aluminium end brackets. The rotor is of the two-piece 8-pole construction, being supported at each end by ball-bearings in the end brackets. The rotor carries a field winding connected to two cupro-nickel slip rings; the brush gear is mounted in the slip ring end bracket; incidentally, the rotor can be *driven in either direction of rotation*.

Alternator Output Control. The output is controlled to suit the battery requirements by means of a vibrating-contact voltage regulator in the field circuit. It is not necessary to employ an *external current limiting device* to control the output of the alternator, its reactance being such that the current is limited to 60–65 amps. when cold and 52–57 amps. when hot, at all speeds up to 11,000 r.p.m. (of the generator rotor).

The generator is supplied with its special design of regulator, or control box and also an *isolating relay* to switch off the alternator field and voltage regulator windings when the ignition is switched off.

The Control Box (Fig. 71). This unit contains a vibrating-contact regulator which controls the alternator's field current and, therefore, the output voltage. A transistor has been incorporated since the field current of an alternator is more inductive than that of a dynamo, so that any direct make-and-break action in this circuit by means of vibrating contacts would cause contact burning and oxidizing. By using a transistor the contacts are

Fig. 71. The Lucas generator control unit, with cover removed

subject only to a small non-inductive pilot current, while control of the main field current is effected by the transistor. It is an inherent advantage in the use of transistors that their current amplifier characteristics can be utilized to control, by means of the vibrating-contact voltage regulators, higher values of field current than previously had been possible.

Basically, the control box unit consists of an electro-magnetic vibrating contact type regulator having its operating coil connected across the D.C. output terminals of the alternator stator, and a pair of normally closed contacts in the base circuit of a transistor. Internal heating of the transistor is dissipated in an aluminium heat sink on which it is mounted. The transistor is protected from inductive voltage surges by a field discharge diode through which the stored energy of the field system is dissipated. Both transistor and diode are exposed to air cooling. The regulator carries a frequency winding to increase the rate of vibration of its armature, thus ensuring an adequately smooth and steady charging current. A base current limiting resistor is connected in series with the regulator contacts.

The control box does not incorporate either a cut-out or a current regulator. As previously stated a cut-out is not required because the rectifier diodes in the alternator prevent reverse currents from flowing.

Operation of the System. Referring to the circuit diagram given in Fig. 72, when the battery is switched on, the battery voltage is applied to both the voltage regulator winding and transistor. Assuming the usual positive to negative current flow a pilot current flows from the battery positive terminal through the emitter E, its base B, the base resistor, regulator contacts and frame—all shown on the right, in Fig. 72—and thence returns to the battery negative terminal. The positive pole of the battery is shown, earthed in Fig. 72. The transistor action results in this base current causing the alternator field current to flow from the emitter E, through the collector C, frequency coil and terminal F to the alternator rotor winding.

When the engine is started, the alternator output voltage increases until the regulator contacts open, thus interrupting the

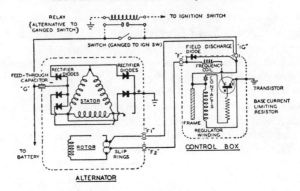

Fig. 72. Circuit diagram of Lucas generator and control unit

base circuit current and, again through the transistor operation, the field current is reduced to a very low value. The alternator voltage therefore drops and allows the contacts to close again. This cycle is repeated at a sufficiently high frequency to maintain the alternator voltage at the predetermined value.

Temperature Effect. The Lucas generator, as controlled by a vibrating-contact voltage regulator, provides a greater current output when cold than when hot. Thus, at a generator speed of 2,000 r.p.m., the cold and hot currents are about 47 and 43 amps., respectively. At 3,000 r.p.m. these values increase to 57 and 50·5 amps. and at 4,500 r.p.m., the respective values are 60·5 and 54·5 amps.

Delco-Remy A.C. Generators. The earlier A.C. generators used on vehicles having greater electrical loadings than standard types fitted with D.C. dynamos, used a special type of power rectifier of the multi-plate pattern mounted away from the generator, to provide direct current for battery purposes. The recent generators—which were among the first of the automobile type—employ a somewhat similar arrangement of diodes and voltage regulator of the transistorized kind to that described earlier. The generators are designed, in regard to their pulley and mounting, to be interchangeable with the standard dynamos.

THE CHARGING SYSTEM 121

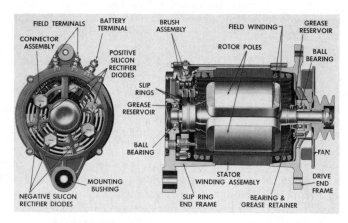

FIG. 73. The Delco-Remy A.C. generator, in end and side views

Fig. 73 shows the 12-volt passenger car generator in end and sectional views, the principal components being indicated by the lettering. The alternator employs six silicon rectifier diodes, of relatively small dimensions but of large current flow capacity, mounted on the end cover. The field and battery terminals are shown at the top of the cover.

The generator rotor is mounted on ball-bearings at each end; each bearing has a grease reservoir, providing for efficient lubrication over a very long period. The two brushes which bear on the slip rings and carry current to the field winding are extra long in order to provide long periods of service.

The end cover is ventilated and a large diameter fan, attached to the twin groove driving belt pulley provides the internal cooling air.

As with the previously described alternator, the regulator uses a single transistor and a diode. The transistor works in conjunction with a conventional voltage control unit having a vibrating contact. The diode reduces arcing at the contacts by dissipating the energy created in the field windings when the contacts separate.

The generator model shown in Fig. 73 gives a current of

25 amps. at fast idling speeds and a maximum of 60 amps. Since the operation and circuit arrangement is similar to those described earlier, it is unnecessary to repeat this information.

The Chrysler A.C. Generator. Fitted as standard equipment to cars made by the Chrysler Corporation, these generators provide higher current outputs from compact units, enabling battery charging to be done when the engine is idling and giving about three times the useful life of a D.C. dynamo. A further advantage is that the bearings are self-lubricating over their lifetime.

Fig. 74 illustrates the generator in cross-section, front and rear end views. The two aluminium end housings are held together by three through bolts and the stator assembly is sandwiched between them. The housings are vented at each end and around their outer diameter. Cooling fans are provided at *both ends* of the generator: the cool air is drawn in through the ends and delivered over the stator windings through the outer diameter vents. The rotor consists of a 'doughnut' shaped field coil encased between two six-fingered, overlapping sections which form the pole pieces, so that an effective 12-pole rotating magnetic field is produced. The ends of the field are connected to slip rings. The battery is connected to the field windings through the voltage regulator, the brushes and slip rings.

The rotor has a ball-bearing at the driver end and a roller-bearing at the opposite end. Both bearings are of the grease-packed kind, requiring no periodic lubrication attention.

Six silicon diode rectifiers are mounted in the end cover. Three of these rectifiers have negative polarity cases and are pressed into the die-cast aluminium end housing. These rectifiers are on the earthed side of the electrical system. The three remaining rectifiers have positive polarity cases and are pressed into a die-cast aluminium heat sink, which is electrically insulated from the end housing but has sufficient area to absorb the heat from the rectifiers.

The generator is used with a vibrating contact type voltage regulator and the electrical system is similar in principle to those described previously.

THE CHARGING SYSTEM

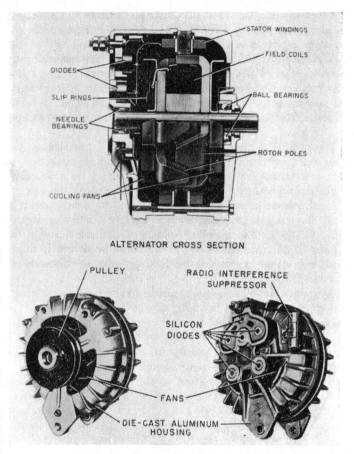

Fig. 74. The Chrysler A.C. generator

The generator output is 28 amps. at 14·6 volts, at 1,250 engine r.p.m. and 33 amps. at 15 volts at 2,200 engine r.p.m.

The field current is 2·38–2·75 amps. at 12 volts and 2·97–3·43 amps. at 15 volts, at 70° F, and 1,750 r.p.m. engine speed.

Commercial A.C. Generators. As mentioned previously, this type of generator is of particular importance for passenger vehicles, where heavy loads and low average engine speeds are the rule. Apart from the other models mentioned earlier, the C.A.V. 24-volt A.C. generator is of special interest. Of 7-in. diameter it has a self-limiting output of 60 amps. at 27·5 volts. The cutting-in speed is 580 r.p.m. With a drive ratio of 2·3:1 the output at normal idling speeds of 350–400 r.p.m. is 19 to 25 amps. With a drive ratio of 2·5:1 the output goes up to 29·5 to 35 amps.

The rotor has two six-fingered cup-shaped portions on the main shaft, forming a 12-pole unit. A single coil is wound around the shaft and is enveloped by the 12 poles of the rotor. The excitation current to the field coil is supplied from the battery through carbon brushes and slip rings; as the field current is small no arcing occurs, even at high speeds. Ventilation is provided by an inbuilt fan which directs cooling air through the interior of the machine. The generator can be driven in either direction, being unaffected by field polarity or direction of rotation of the rotor.

A separate dry-plate selenium rectifier is used and two alternative types of control units, using conventional vibrating type or transistor-type regulators, are available.

The generator operates up to 4,500 r.p.m. (generator speed). It absorbs 3 b.h.p. at 2,200 r.p.m. and weighs 37·5 lb. The rectifier weighs 20 lb. and control unit 6·5 lb.

In the later C.A.V. alternators, covering the 12-, 24- and 32-volt range, the maximum outputs range from about 23 amps. at 5,000 r.p.m. to 55 amps. at the same speed. In these, rectification is by means of built-in silicon diodes, mounted on heat sinks at the slip ring end, connected between the stator windings and the output terminals. They are known as the AC5 type alternators.

CHAPTER 4

IGNITION SYSTEM PRINCIPLES

In order to understand the principle of the ignition system it is necessary to be conversant with the theory of the petrol engine, in so far as it concerns the ignition of the air-petrol charge. It will here be assumed that the reader is familiar with the four-cycle and two-cycle operations employed in petrol engines.*

The purpose of the ignition system is to supply a spark within the cylinder towards the end of the compression stroke, to ignite the compressed charge of turbulent air and petrol vapour.

Referring to Fig. 75, in which the lower diagram shows the piston positions within the cylinder at the beginning c of the compression stroke and end b of this same stroke, the degree of compression is denoted by the ratio of the cylinder plus clearance volume, namely, Oc to the clearance volume Ob and in modern automobile engines, this ratio varies, in different engines, from about 7·5:1 to 10·5:1 in high compression engines. The compression ratio has an important influence upon the ignition requirements of an engine, as will be explained later.

The efficiency and performance of an engine are to a large extent governed by the exact moment at which the ignition spark occurs. This can be demonstrated on the pressure-volume diagram of the ideal four-cycle operation, shown in the upper diagram, in Fig. 75, which shows the compression and heat expansion or 'firing' strokes. Briefly, the air-petrol mixture is taken into the cylinder at C, at atmospheric pressure $C'C$, and then compressed along the curve CB, to the compression pressure $B'B$. When at the end of the compression stroke, namely, at B, the charge is ignited by an electric spark and the pressure, due to the combustion energy, rises suddenly to a maximum value of $B'A$ along BA. Then, the compressed gases

---

* For fuller information, Motor Manual No. 1, *Automobile Engines*, may be consulted.

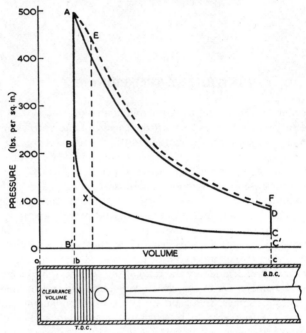

Fig. 75. Effect of ignition point on theoretical power output

force the piston along its working or expansion stroke AD to D, after which the pressure falls along DC to atmospheric pressure at C.

In these theoretical considerations the ignition spark occurs at B, giving an instantaneous increase of pressure from $B'B$ to $B'A$. The area of the diagram $ABCD$ can be shown to represent the amount of work done per cycle by the piston and, therefore, the larger this area for a given amount of air-petrol charge, the greater will be the power output.

Consider, next, the case in which the ignition occurs at a point X, after the piston reaches the end of its compression stroke. The pressure will rise suddenly from X to a value E, which will be less than that at A, since the compression pressure when the spark occurs is at a lower value than at B. The pressure-volume diagram will then be *XEFC* and its area will be appreciably less than that of *ABCD*. Therefore, the effect of the late ignition will be a reduction in work or power output.

It can readily be shown that when the ignition occurs at *X* before the end of the compression stroke the result will be somewhat similar, leaving a smaller net work area. Not only is there *a loss of power* but also an increase in fuel consumption when the ignition occurrence, or *timing*, is different from that indicated at *B*.

Effect of Ignition. So far these considerations have been confined to the ideal cycle of operations, which assumes an instantaneous rise of pressure, due to heat addition—by combustion of the charge—along *BA*.

In practice, however, the charge does not burn instantaneously but requires a definite, but small period of time to attain its maximum temperature and pressure, so that the pressure-volume diagram becomes modified.

The ignition delay between the occurrence of the spark and attainment of maximum pressure depends upon the mean flame velocity, the compression pressure, the engine speed, the design of the combustion chamber, position of the sparking plug and certain other minor factors, which are outside the scope of this volume.*

The *practical effect of this delay period* is allowed for in petrol engines by arranging for the spark to occur at some point before the piston has reached the end of its compression stroke, usually when the crank of the piston assembly is at an angle of 30° to 45° before its top centre position. This angle is known as the *ignition advance angle*. Its actual value, for maximum output and minimum fuel consumption, is usually found experimentally for any particular engine.

The importance of *correctly timing the ignition* of an engine is illustrated by the results shown by the pressure-volume diagrams shown in Fig. 76 and which were obtained by means of an indicator, from an actual engine.

Diagram (A) was obtained when the spark occurred too late

* For fuller information refer to *Modern Petrol Engines*, A. W. Judge (Chapman & Hall Ltd., London).

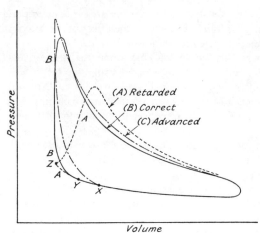

Fig. 76. Effect of ignition point, or timing, as shown on actual indicator diagrams

or *retarded*; Diagram (B), when the ignition timing was *correct* and Diagram (C), when the spark occurred too early, i.e. too *advanced*. It will be seen that the full line diagram (B) has the greatest area, thus giving the greatest power and fuel economy. In this case the spark occurs at Y.

In the case of Diagram (C), where the ignition spark occurs too early at the point X in the compression stroke, the pressure commences to rise above the compression line, along XB and reaches a higher peak pressure—due to detonation effects—than (B). Its expansion line, however, is mostly above that of (B), but in practice its net area is less.

For the retarded ignition Diagram (A) the spark occurs at Z and the pressure rises along ZA to a lower peak value than in the case of (B) and (C). The expansion line, however, lies above that of the other two diagrams, but the net area is less than for these diagrams.

The Ignition Agent. A compressed charge of hydrocarbon fuel, e.g. petrol or Diesel oil, and air in the correct proportions for combustion, can be ignited by any device which is at a sufficiently high temperature above that of the charge, i.e. above the fuel's self-ignition temperature. In the early days of the motor-car, a gas flame in a tube, i.e. a flame tube, was exposed, by means of a slide valve, to the compressed charge at about the correct moment, for ignition purposes, but this method was superseded by the electric spark systems, which included the trembler, or Ruhmkorff induction coil that produced a shower of sparks, and the 'break-spark method', depicted in Fig. 77.[*][†]

In this method, which was used on some European and American engines, a battery, X, supplied current through an inductive coil Y to an insulated plug in the engine cylinder head. A lever B, worked by a rod running on a cam E, caused the contacts C to open once every two revolutions of the crankshaft. Normally the plain cylindrical face of the cam E, allowed the fixed and moving contacts C to be pressed together by a spring, so that current from the battery X flowed through the coil Y,

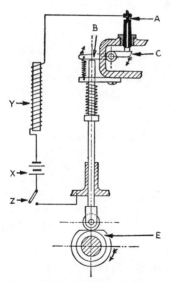

FIG. 77. The make-break spark ignition system

[*] *Petrol Motors and Motor Cars*, F. Strickland, 1914 (Griffin Co.).
[†] Vide page 133.

the plug A and contacts to the metal of the engine; the other pole of the battery was earthed to the engine, to complete the circuit.

When the contacts C were separated by movement of the rocking lever B, the self-induction of the coil Y caused a spark to pass between the contacts. The energy of this spark was proportional to the self-induction of the circuit, at the moment of the contacts 'breaking' and to the square of the current. Thus:

$$\text{Heat energy due to 'break'} = \frac{C^2 L}{2}$$

where C = current and L = self-induction at 'break'.

For practical reasons this ignition system did not survive, being superseded by the previously mentioned trembler coil method which in its earliest application at the beginning of this century, operated from a battery and, later, employed a low-tension magneto as the current generator. It is interesting to note that the original Ford Model T employed this method. In this case current was generated by sixteen magnets mounted around the rim of the engine flywheel and sixteen coils arranged in series to form the stator. This four-cylinder engine employed a separate trembler coil connected to its own sparking plug. A distributor device directed the current to the primary winding of each induction coil, in turn, thus causing the secondary coil to produce its stream of high-tension sparks over the predetermined ignition period.

While this method worked fairly well, if maintained in first-class order throughout, it was not altogether reliable; pitting of the contacts was a typical cause of failure. The trembler coil method was eventually replaced by the high-tension Bosch magneto and its variants and, in more recent times entirely—so far as automobile engines are concerned—by the coil-ignition system. The magneto and coil-ignition systems are described later.

The High-Voltage Spark. Irrespective of the source of electrical supply, when a sufficiently high voltage is applied to a circuit containing a pair of metal sparking points, and these points are arranged at a certain distance apart, a high-voltage

IGNITION SYSTEM PRINCIPLES

spark will flash from one point to the other. As employed in the case of the automobile sparking plug, the sudden increase of voltage generated in the secondary coil of a type of transformer produces an intense spark between the central insulated and outer metal or earthed electrode of the plug, which is sufficient to commence the ignition of the compressed air-petrol charge; the combustion flame spreads rapidly from the plug electrodes through the combustion chamber, so that all of the petrol vapour is burnt.

Before proceeding to the practical consideration of ignition systems it may be of interest to discuss, briefly, some of the theoretical and experimental aspects of the subject.

The Electronic Explanation. For a spark discharge to occur between two metal sparking points the surrounding air or gas must become a conducting medium instead of its normal insulating state. This process of conduction under high voltage conditions can be brought about by an electronic process known as *ionization*. To appreciate this process it is necessary to understand the electronic theory of matter, whereby it is considered that an *atom* of any element consists of a nucleus having a positive electric charge, surrounded by a number of negative charges, known as *electrons*. The latter may be regarded as revolving around the nucleus and their number and arrangement determine the type of atom, or element. A molecule, in its neutral state has equal positive and negative charges. The electrons, which can under certain conditions have a separate, or free state, are capable, by impact action, of changing the polarity of the molecule; thus, by the loss of an electron, the molecule acquires a positive state, and is known as a *positive ion*. The gain of an electron produces a *negative ion*. It is these ions which carry an electric charge.

In the case of a pair of sparking points surrounded by a gas, when a difference of potential exists an electric field occurs, any stray negative ions will be attracted to the positive electrode, or sparking point and positive ions to the negative electrode. As the potential difference between the sparking points increases more and more of these ions are created until, eventually, the gap between the points becomes ionized until at a certain potential difference the gap becomes a conductor, so that an electric discharge, or spark occurs; this is sometimes known as a *disruptive discharge*.

It follows that before a disruptive discharge can occur there must be an ionization in the gas between the sparking points. With a steady voltage (potential difference) across the points an appreciably lower voltage is required to produce a spark than when the voltage is of the surging or impulsive type that is produced by the method of a sudden break in the primary circuit of a transformer. The ratio of the impulse to steady voltage is known as the *impulse ratio* of the sparking gap. This ratio is usually high at starting, owing to scarcity of ions, but under normal working conditions in the combustion chamber, where an ample supply of ions usually exists, the ratio approaches unity.

It is important, in connection with sparking plug voltage tests to ensure that there is sufficient ionization at the sparking electrodes—a condition which does not always exist with ignition testing apparatus. It is for this reason that voltage measurements are often made with an artificial ionization method, known as the *three-point gap*; this is described later.

Production of the High-Voltage Sparks. The principle of the method employed to produce ignition sparks by the trembler coil, magneto or ignition coil, is that of a primary circuit consisting of relatively few turns of a thicker insulated wire, surrounded by a secondary coil having a relatively high number of turns of thinner insulated wire.

The primary coil is supplied with a low voltage current, of about 6 to 12 volts, and is provided with a device for breaking the current flow, quickly and regularly. This sudden break in the primary current causes a high-voltage impulse to occur in the secondary coil circuit; it is this impulse that is used to produce the spark at the sparking plug electrodes.

It may be mentioned that the voltage produced in the secondary circuit is roughly proportional to the ratio of secondary to primary coil turns, multiplied by the primary voltage. The actual current produced in the secondary circuit is, relatively, very small.

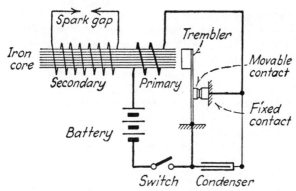

Fig. 78. Principle of the Ruhmkorff or induction coil

The type of compound coil or transformer referred to is based upon the *induction, or Ruhmkorff coil*, the principle of which is shown in Fig. 78. In this illustration, for convenience, the primary coil, consisting of a few turns of wire, is shown on the right, surrounding an iron core. The secondary winding, of a relatively large number of turns on the same core, is indicated on the left. In practice one coil would be wound around the other, as is explained more fully, later.

The primary circuit has two contacts, one being fixed and the other, mounted on a strip of spring steel, is movable. At the upper end of the spring is a short soft iron cylinder trembler. When the battery switch is closed, current flows around the primary coil and the closed contacts, back to the battery. The effect of the current in the primary coil is to create a strong magnetic field, which causes a magnetic attraction between the core and the trembler; the trembler thus moves to the left, causing the movable contact to separate from the fixed contact, thus breaking the primary circuit. The current is therefore switched off and the magnetic field collapses, so that the trembler, under the influence of its spring support, is moved back; then the movable contact closes on to the fixed contact. The cycle is then repeated, so that the trembler rapidly completes and breaks the primary circuit, all the time the battery switch is on. This causes the magnetic field to build up and collapse rapidly. Due to the time taken to break the primary current being much less than that to build it up, the voltage at the 'break' is much greater than that at the 'make' of the circuit; this delay in the increase of the current at the 'make' is due to the inductance or 'electric inertia' of the circuit.

The effect of the rapid build up and collapse of the primary circuit magnetic field is to cause an electromotive force of several thousand volts to occur in the secondary circuit every time the primary circuit is broken in an actual coil. If, as shown, the ends of the secondary coil are connected to a spark gap of suitable dimensions, a high voltage spark will flash over the gap every time the primary circuit is broken.

When applied to the ignition of a petrol engine mixture, an engine-driven cam was used to control the timing of the spark, or sparks, in relation to the position of the engine piston near the completion of the compression stroke.

The modern magneto and coil-ignition systems operate on a similar electrical principle but, instead of using the trembler method which can produce a continuous and rapid series of sparks, a single, but oscillatory spark is obtained at the sparking points.

The Condenser. The condenser, shown in Fig. 78, performs two important functions, which may be better understood by first considering the contact breaker unit *without* this condenser. When the contact points separate, the current in the primary circuit tends to continue to flow across the gap so that a spark discharge will occur as the contact points separate. This will soon result in pitting of the mating surfaces, with ultimate breakdown of the primary circuit. Further, if this flow of current during the separation of the contacts is allowed to persist, the current would be reduced more gradually, with the result that the magnetic field would collapse more slowly and therefore produce a lower induced voltage in the high tension circuit.

If, however, a condenser of suitable capacity, namely about ·15 to ·35 microfarad, is shunted across the contact breaker points—as shown in Fig. 78—the condenser will absorb this current until it becomes charged, so that as the points separate there will be practically no current flow, and therefore no significant sparking, so that the contact points will remain comparatively clean over long periods of service.

Again, this prevention of current flow during the point-opening phase, enables the magnetic field to collapse more suddenly, thus producing a higher voltage in the secondary circuit, for the ignition spark. Further information on ignition condensers is given in the chapter that follows.

Principle of Magneto and Coil Ignition Systems. The principle of both of these ignition systems is illustrated in Fig. 79, which shows both the primary, or low tension and the secondary or high tension circuits of a battery-supplied system. If the current generator of the magneto is substituted for that of the battery, it will not affect the principle since both are low voltage current supply systems.

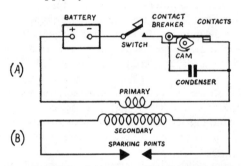

FIG. 79. Principle of magneto and coil ignition systems

The primary and secondary circuits are shown separately, for convenience of explanation, but, in practice the two coils would be wound concentrically, one over the other and a common soft-iron core would be used. This arrangement is that of the *high tension coil*.

Operation of Circuits. The primary circuit has an automatic device for making and breaking the circuit. This device comprises a light spring-controlled type of movable contact which is lifted once every two revolutions, in the case of a four-cycle engine, by means of a small rotating cam, driven at one-half engine speed. For two-cycle engines the cam runs at engine speed, to produce one spark in the complete circuit, every engine revolution. A condenser is connected in parallel across the contact breaker points.

Assuming the battery is switched on and the engine is operating, low voltage current will flow through the primary circuit all the time the contacts are closed, but when the cam rotates to the position where its projecting face lifts the spring arm of the upper contact, the contacts separate, to break the primary circuit.

While the primary circuit contacts are closed, a magnetic field is created through the common core and around the primary and secondary coils. However, owing to the self-induction of the primary circuit the magnetic field does not build up to its full strength at once. This is due to the fact that self-induction in a circuit produces a counter-electromotive force, which opposes the battery current, so that this must be overcome by the battery voltage. This delay in building up of the magnetic field is termed the 'build-up time' or 'period'; it is an important factor at the higher engine speeds, rendering efficient spark production more difficult.

When the contact breaker points separate, the magnetic energy stored in the field is released and this induces a voltage in the secondary winding, its value being approximately in proportion to the ratio of secondary to primary coil winding number of turns.

The secondary circuit has a certain capacitance in the windings, cables, etc., which is equivalent to that of a condenser in the secondary circuit, in parallel with the spark gap. Due to the collapse of the magnetic field, the increasing voltage in the secondary circuit charges this (equivalent) condenser to a sufficiently high value to cause the voltage to produce a spark at the sparking points.

The spark thus caused is known as the *capacitance spark component*.

Example. If it is assumed that the breakdown voltage causing a spark to bridge the spark gap is 10,000 volts and the circuit capacitance is 50×10^{-12} farad, then the energy released by the initial breakdown will be $\tfrac{1}{2}C_1V^2$, where C_1 is the capacitance and V the breakdown voltage. Substituting for C and V the assumed values the result will be as follows:

Energy release $= \tfrac{1}{2}C_1V^2 = \tfrac{1}{2} \times 50 \times 10^{-12} \times 10{,}000^2$
$\phantom{\text{Energy release }= \tfrac{1}{2}C_1V^2 }= \cdot 0025$ joule

The Inductance Component. It can be shown that the capacitance energy is only a fraction of the inductive energy stored in the magnetic field, so that it is necessary to take this inductance component into account, also.

The inductive energy is given by the expression $\tfrac{1}{2}LC^2$ where L is the inductance and C the current.

With the high-tension coils used in automobile ignition circuits, the inductive energy at low engine speeds and when the secondary voltage is a maximum, may attain the value ·04 joule for each spark produced, in a coil ignition system. The effect of increase in engine speed is to lower the inductive energy.

The electrical discharge caused when an inductive circuit is broken down is known as the *inductance spark* and this spark contributes the greater energy to the spark discharge.

It was mentioned earlier that ionization exists in sparking plug

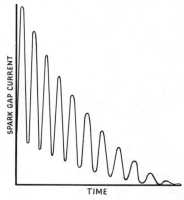

FIG. 80. Showing oscillatory nature of the spark gap current

gaps, this effect causing the resistance of the gap to be smaller, so that the value of the *spark current* depends upon the inductance, while the *spark duration period* depends upon the secondary coil winding resistance. The inductive energy discharge is of an oscillatory nature, while the capacitance discharge is unidirectional, so that the net effect of the two spark components is to produce a spark current which is not only of an alternating nature but of a unidirectional oscillatory character as depicted in Fig. 80.

Some Spark Component Considerations. While the capacity component of the spark energy is so much lower than that of the inductive energy, it has been shown that in the case of a petrol engine having clean sparking plugs, adjusted to the correct gaps, namely, between ·012 to ·015 in., and with the correct proportion of air to fuel supplied by the carburettor, namely, about 15 of air (by volume) to 1 of petrol, the capacitance component will supply all the energy required to produce the ignition spark. It is thus possible, and practicable, to use ignition devices which produce relatively small discharge energy, but with rapid rise of spark voltage.

In this connection, the results of studies made with the cathode ray oscillograph have revealed the fact that automobile ignition appliances produce secondary circuit voltage of *extremely rapid rates of rise*, namely, of the order, 40 to 50 million volts per second for coil ignition systems and 300·to 400 million volts per second for magneto ignition systems.

The function of the inductance component, with its relatively high energy content is not so effective as might be anticipated, since it has been shown, experimentally, that only a short initial portion of this component is required for ignition purposes; further, once ignition has been initiated, the balance of the spark energy dissipation has no beneficial effect.

It has been suggested that the inductance component may give a quicker rise in pressure in the case of weak-in-fuel mixtures—which have a relatively low rate of pressure rise; further this component may be helpful in assisting volatization of the mixture fuel in the case of cold engines. Again, it is possible that the surplus energy of the inductance component may be useful in the case of a sparking plug having dirty insulation around the sparking electrode, which would lower the resistance so that if too little energy were available, misfiring might occur. If, however, a greater amount of energy were available—as with the inductive component—regular ignition across the sparking plug electrodes would take place. In the case of coil ignition systems, this reserve energy would be available at low speeds while for magnetos, it would be effective at high speeds. If, however, too much energy is used in the latter case, the plug points would burn away more rapidly. To obviate this risk certain magnetos incorporating devices to reduce the spark energy at the higher speeds have been used, in the past.

Spark Energies. The heat energy that is necessary to ignite an air-petrol mixture in the combustion chamber of an engine, is dependent upon a number of factors of which the more important are: (1) The ratio of air-to-fuel. (2) The compression pressure. (3) The temperature of the mixture. (4) The degree of turbulence in the mixture. The results of experimental work by B. Lewis of the U.S.A. have shown that there is a definite relationship between the minimum sparking energy and the length of the spark gap. While the experiments were made with free and glass-flanged sparking electrodes, the results are applicable, generally, to petrol engine conditions.

It was shown that with the electrodes, ·01 in. apart, the minimum spark energy was about ·008 joules. When ·04 in. apart, ·003 joules. When ·08 in. apart, ·006 joules and, thereafter for gaps, from ·09 in. to ·16 in. the energy was constant at about ·005 joules. The critical gap, therefore, was ·09 in. and if the gap was smaller than this value, the spark energy increased progressively with the reduction of the gap.

Under petrol engine combustion conditions, using the correct air-petrol ratio (about 15:1), maximum pressure and temperature and full throttle, the spark gaps, or quenching distances, need be very small for ignition purposes, namely, ·010 to ·016 in.

For modern engines with compression ratios of 7·5:1 to 10·5:1, the maximum gap need not exceed ·025 in., but for part throttle operating conditions—which usually correspond to the greater part of normal car engine usage—somewhat larger gaps are generally required for reliable ignition and slow running; it is for these reasons that sparking plug gaps of ·030 to ·035 in. are used in many instances.

It has been stated that the minimum spark energy for ignition purposes is of the order ·005 joules, whereas the total of the capacitance and inductive energies of typical coil ignition high tension circuits was of the order ·04 to ·045 joules. It follows that the available sparking energy is from eight to nine times that required for ignition purposes in the present example, which is, however, a typical one. More generally, modern electrical equipment provides energies of ·04 to ·08 joules, so that the surplus energy is even greater than that of the example chosen.

Another important factor is that of the *spark voltage*, since it has been demonstrated that as this voltage is increased, the minimum spark energy for ignition is reduced, but at a more rapid rate. Thus, in a typical case, for a somewhat richer air-petrol mixture of 12·5 : 1 than the correct mixture for complete combustion of the fuel, the spark energy required to ignite this charge was ·005 joules at 5,000 volts, ·002 joules at 6,000 volts and ·0007 joules at 7,000 volts.

Effect of Mixture Strength. The results of tests made by Paterson and Campbell on the minimum spark energies required to ignite air-petrol mixtures of different ratios, showed that this energy is a minimum for a mixture consisting of about 10 per cent petrol and is greater for other mixtures both weaker and richer than this mixture.

Fig. 81 shows the results of these tests, when using a sparking voltage of 5,280 over a range of mixture strengths. It will be observed that the minimum energy value is ·0035 joules for a mixture ratio of 9·75 : 1, while for the approximately correct mixture of 14·5 : 1 the value is ·008 joules.

It may be mentioned that these laboratory values were subsequently confirmed by tests upon an aircraft engine, the spark energy being reduced by an external control until misfiring occurred. It was found that the minimum energy at which the engine would operate satisfactorily was ·0014 joules.

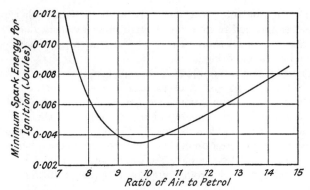

Fig. 81. Influence of mixture strength on spark energy

Factors Influencing Sparking Voltage. The principal factors concerned in ignition high tension circuits may be summarized briefly, as follows:

Charge Pressure. The effect of charge pressure is to increase the density of the charge and, since the voltage required to bridge a given gap, surrounded by a gas, increases with the gas density, as the compression ratio of an engine is raised, so will the ignition voltage, at a given gas temperature, have to be increased.

Throttle Opening. It is only at full throttle opening that the maximum charge density is obtained. As the throttle of the carburettor is closed, progressively, so is the density of the charge drawn into the cylinder reduced, and in consequence, the voltage needed to produce a spark at the sparking plug is reduced.

Charge Temperature. Increase of charge temperature, due to atmospheric conditions, has the effect of reducing the charge density so that a lower spark voltage is required.

Intensity of the Ignition Spark. There has frequently, in the history of the petrol engine using electric ignition, been some misconception concerning the effect of spark intensity upon engine running and performance. It has been claimed that more power is obtained from a 'fat' or high intensity spark than from the less intense or normal spark and spark 'intensifiers' have been offered to the motoring public.

The results of tests made, over half a century ago, by Prof. W. Watson, using sparks of varying intensities from magnetos and coil-ignition units showed, conclusively, that, provided the spark was sufficiently intense to ignite the mixture, whether weak or rich, under full throttle conditions, any further increase in intensity had no effect upon the power output; on the other hand, the more intense sparks have the disadvantage of burning away the sparking plug electrodes more quickly than with the normal spark.

Plug Electrode Temperature. As the sparking plug electrode temperature increases, after an engine has been started from the cold, so the voltage needed to produce the ignition spark diminishes; this may be due to the existence of a film of heated gas on the electrodes at a density lower than that of the charge.

Another effect of temperature increase is to *reduce the insulation resistance* of the sparking plug. Thus, in the case of a steatite insulator plug having an insulation resistance of 20 megohms at 350° C the resistance fell rapidly to 1·5 megohms at 500° C.

When an engine is started from the cold the lower charge and plug electrode temperatures require a higher voltage than when the engine has attained its normal operating temperature. When, however, the engine is running under the larger throttle openings the increased charge density practically compensates the plug electrode temperature effect upon the sparking voltage.

The temperature of the plug electrodes can be to a large extent controlled by plug design, a subject which is discussed later in this volume.

Turbulence of the Charge. A higher voltage is needed to bridge a given plug gap in a combustion chamber promoting turbulence of the incoming and compressed charge. Some idea of this effect can be gleaned from the results of experiments made with two sparking plugs that were mounted on a rotatable arm,

IGNITION SYSTEM PRINCIPLES 143

in air, provision being made to measure the sparking voltages at various speeds of rotation. It was found that when at rest each plug required a sparking voltage of 2,900. At plug linear speeds of 10, 20, 30 and 40 ft. per sec. the peak voltages required were, respectively, 3,500, 4,000, 4,300 and 4,500.

Measurement of Sparking Voltage. For the purpose of testing ignition equipment on the test bench it is necessary to reproduce, as nearly as possible, the actual conditions in the engine cylinder. In this connection *it is unsatisfactory* to connect a sparking plug to the ignition unit, in air, since at atmospheric pressure, a relatively small voltage is required to produce a spark across the plug electrodes. Neither is it desirable to use a longer air gap, e.g. from 0·3 to 0·5 in. These methods do not provide for proper ionization which, as explained earlier, is an essential factor.

The usual method for air tests of sparking voltages is to provide a means for ionization and also to simulate the conditions of sparking plug fouling by shunting a resistance of about 200,000 to 400,000 ohms across the spark gap.

The *production of an ionized sparking gap* can be effected in two ways, namely (1) By making the electrodes of relatively large dimensions so that there is always the probability of the essential

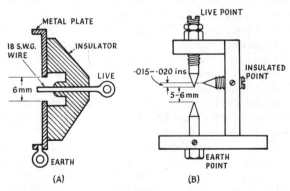

Fig. 82. Types of spark gap for testing sparking voltages of ignition systems

ions being present in the vicinity of the gap or (2) By providing for artificial ionization at the gap. The latter method is the one more generally favoured for test purposes and it can be made in the two forms illustrated in Fig. 82.

The Annular Spark Gap. This is essentially a device more suitable for routine or duration test purposes, since the electrodes burn away at a much slower rate than for the three-point gap. A convenient annular spark gap for normal testing purposes consists of a nickel or alloy rod of about 18 S.W.G. (·048 in.) arranged at the centre of a ¼-in. diameter hole in a steel plate of about 18 S.W.G. The edges of the hole should be rounded off. With such a gap, sparking will occur regularly, in air, at about 8,000 volts. For a voltage of 10,000, the central electrode should be of 16 S.W.G. (·064 in.) and the hole of ·345 in. diameter.

The Three-point Gap. This has a pair of sharp-pointed electrodes or points, one of which is connected to the high voltage supply, while the other is earthed. The third point is insulated from the other two points and is separated from the *live point* by a gap of only a few thousandths of an inch. It is important that this point should be located slightly out-of-line with the live point, by about $\frac{1}{64}$ in. Due to the capacitance of the third point a tiny spark occurs between this point and the live point just before the voltage rises to its maximum value; this spark is concerned with the production of the ionization conditions and ensures regular sparking from the live to earth points at a much lower voltage than for a two-point spark gap, without ionization.

The distance between the sparking points, to produce regular sparking with voltages of 8,000–10,000 should be 5–6 mm.; in practice, the discharge voltage depends upon the voltage wave characteristics; some wave forms, e.g. in magnetos at high speeds, steepen, to give voltages in excess of 10,000.

It will be apparent that, with its sharp sparking points, the three-point test gap will wear away quickly; for this reason, the points are sometimes replaced by balls of hardened steel, of ¼ in. diameter. With the three-ball test gap it is found that for a given applied voltage the sparking gap is appreciably less than for the three-point arrangement; in one particular series of ignition spark tests it was found that whereas with the three-point gap a sparking voltage of 10,500 was required for a gap of 5 mm., with the three-ball gap, a voltage of 10,000 was required for a gap of only 2·5 mm. It is generally agreed, however, that the three-ball gap is more reliable for gaps below 3 mm.

Using the Three-point Gap. The three-point gap is employed where a number of different voltage tests have to be made on coil-ignition or magneto-ignition equipment. Such tests would include: Determination of minimum engine speed giving regular sparking at fully-retarded and fully-advanced spark timing positions; *endurance test* at maximum speed; *utility test*, i.e. testing with a leakage path of known value placed across the test gap electrodes, to ascertain the upper and lower limits of the speed range giving regular sparking. In this connection the leakage path consists of a variable resistance shunted across the sparking points, to simulate plug insulator leakage due to oily depositions; usually the resistance range is 200,000 to 500,000 ohms.

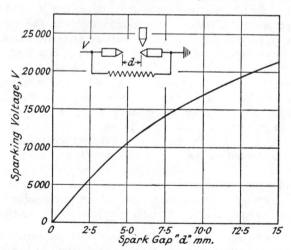

FIG. 83. Sparking voltages for different spark gaps (three-point gap method)

Sparking Gap and Voltage. The results of a series of sparking voltage tests made with the three-point gap, shunted with a resistance to simulate the most unfavourable operating conditions are shown in Fig. 83. From these it will be seen that the sparking voltage increases as the electrode gap is increased, so that for a gap of approximately one-half inch, in air, a voltage of just under 20,000 would be required.

The sparking voltages employed in various engines, of different compression ratio, speed range, combustion chamber design, plug location in combustion chamber and other variants may be as low as 6,000 to 8,000 volts and, in extreme instances, as high as 25,000 volts; the latter voltage would correspond to very high engine speeds, with high compression ratios, as for racing car engines.

Example of Ignition Requirements. From both theoretical and experimental considerations it has been shown that the ignition system must supply the total energy that is needed to statically charge the capacitance of the ignition high tension circuit to a sufficiently high voltage to insure regular ignition of the cylinder charges under the most unfavourable conditions. The capacitance of the high tension circuit includes the capacities of the H.T. secondary coil, leads, sparking plug and distributor.

A typical example, due to Dr. E. A. Watson, is that of a high tension system having a total capacitance of 100×10^{-12} farad and a sparking voltage of 25,000 volts. With this capacitance the ignition system must provide an energy output of ·0031 joules. In order to provide this energy at high engine speeds without exceeding a maximum current of 3·5 amps. at low speeds—since this is about the maximum current the contact breaker points can safely withstand—a 12-volt system would require a primary inductance of about 10 millihenries, i.e. ·01 henry.

Under these conditions sufficient energy would be available under starting conditions, with a low battery, and at all speeds of the engine, assuming the dynamo performance was satisfactory.

Magneto Ignition. It is necessary here to mention the alternative method of ignition spark production, known as the magneto system in which a small engine-driven dynamo or

IGNITION SYSTEM PRINCIPLES 147

generator is employed to supply the low tension current to the primary circuit of its integral low and high tension coil. The current generator may be regarded as a substitute for the battery but, instead of providing a steady current it produces an alternating or fluctuating current in the primary circuit when the contact breaker contacts are closed. It is arranged for the contact breaker points to 'break' or open when the current reaches its maximum value.

For present purposes, therefore, the magneto system may be understood to be the same as that of the coil ignition system hitherto considered, except that the current generator of the magneto replaces the battery. The circuit arrangement is similar to that shown in Fig. 79 if, instead of the battery, the magneto generator be substituted. Thus, it will be seen that the combined low and high tension coils, the contact breaker—which is engine-operated in both cases—and the condenser are the same in each instance. There are certain minor differences in regard to these two systems, but these do not affect the general similarity of their operation in producing ignition sparks.

Since the subject of magnetos, their construction and operation is dealt with later, it is proposed to consider here the performance aspects, in order to compare the merits of the two alternative systems.

Comparison of Coil and Magneto Systems. In the assessment of the merits and demerits of these two systems for automobile engine purposes it should be pointed out that, apart from electrical considerations the question of relative manufacturing costs is of primary importance in these days of mass-production cars, so that as long as an ignition system fulfils its main purpose of providing ignition sparks, at correct intervals and with correct timing, under all engine throttle and speed conditions and, further, will enable the engine to be started from cold, this system, provided that it is fully reliable, is favoured by the car manufacturer.

For special applications, however, the system possessing the best electrical characteristics is adopted, irrespective of initial cost. Thus, for racing engine purposes, where engine speeds up to and often exceeding 10,000 r.p.m. might be used, special ignition systems, including the more recent electronic ones, would be adopted.

In the case of the coil ignition system, the maximum current available in the primary circuit is that of the battery; this is particularly advantageous for engine starting purposes. With the magneto the primary current is zero when the engine is at rest, but the current builds up as the speed is increased and, at a certain designed speed attains its maximum value—which may be appreciably higher than that of the equivalent coil ignition current.

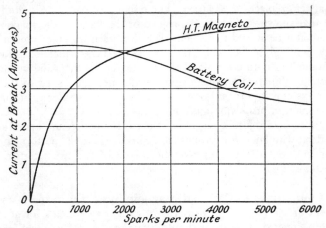

FIG. 84. Comparison of coil and magneto system performances

In this connection the graphs reproduced in Fig. 84, which are based upon the results of tests made by A. P. Young and H. Warren, show the current values at the moment of the 'break' of the contact breakers of equivalent coil and magneto ignition systems. From these curves it will be observed that whereas the coil system gives a maximum current of 4 amps. at zero engine speed, the magneto gives zero current, but as the engine speed increases, the coil system attains its maximum current of 4·1 amps. at 1,000 sparks per min., while the magneto current at this speed is only about 3·25 amps. The coil system current falls

IGNITION SYSTEM PRINCIPLES

away continuously, beyond 1,000 sparks per min., whereas the magneto current increases all the time to about 4·6 amps. at 6,000 s.p.m.; the battery system current, at this speed was about 2·6 amps.

It was shown that the spark energy of the magneto shows a similar advantage over that of the coil system, except at low engine speeds. Thus, while in Fig. 84 the current at 'break' of the magneto and coil system is the same at about 2,000 sparks per minute, the spark energies are equal at a value of ·045 joules for a speed of only 2,000 sparks per min., after which any increase in speed results in an increase in the spark energy of the magneto to a value of ·065 joules at 4,000 sparks per min., and a decrease to ·025 joules for the coil system at the same speed.

It should be pointed out that since the results shown in Fig. 84 were obtained, both the magneto and coil systems have been developed to vary the spark energies at different speeds to suit modern engine requirements, although the general characteristics of the two systems as depicted by the graphs remain.

Sparking Voltages of the Two Systems. A comparison of the voltages developed by a coil system and magneto, in terms of the equivalent spark gap distances and engine speeds, is shown in Fig. 85.

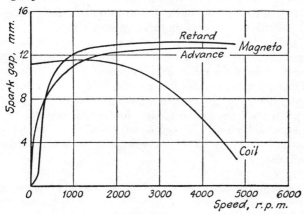

FIG. 85. Effect of speed on coil and magneto system sparking voltages

The equivalent voltages to the spark gaps may be ascertained approximately from Fig. 83, which shows the voltage-gap relationship for a three-point gap test unit.

The voltage characteristics are somewhat similar to those shown, previously, in Fig. 84, since the voltages increase from zero at zero speed to a maximum, in both the ignition timing full advance and retard settings, at the maximum test speed of just under 5,000 r.p.m.

Here, again, the coil system shows a progressive fall of sparking voltage with increase in engine speed beyond the maximum value at about 1,000 r.p.m.—corresponding to the 'fast-idling' engine speed.

The following are, briefly, the principal merits and demerits of the coil and magneto methods of ignition:

Coil Ignition.
(1) Gives a much better spark for starting purposes.
(2) Cheaper to mass-manufacture and to replace the separate component parts, e.g. the coil or the contact breaker parts.
(3) With the magneto the adjustment of the spark timing on the magneto, i.e. advance or retard, causes an alteration in the spark energy, whereas, with the coil system there is no such detrimental effect over the full timing range.
(4) Modern coil ignition units are just as reliable as the magneto. The battery is the only probable unreliable item.
(5) The items of maintenance attention can usually be made more accessible than those of the magneto, except possibly in the case of the camshaft magneto.
(6) For high speed engines it is possible to design a coil ignition system to give the desired sparking voltages at the higher speeds, without the tendency for excess voltages of the magneto; indeed, as mentioned earlier, the tendency is towards a falling off of the voltage towards the maximum speeds.

The principal *disdavantage of the coil ignition system* is its dependence upon the battery. Thus, if the battery has run down, the engine cannot be started or it may crank the engine at too

IGNITION SYSTEM PRINCIPLES

low a speed. In engines of certain transportable equipment and tractors, where battery charging facilities are not available, magnetos must be used.

Magneto Ignition.

(1) In general more reliable, since the design and development of the magneto has resulted in exceptional reliability under all operating conditions. Moreover, it is independent of the battery condition.
(2) Can be adapted more readily for higher engine speed operation.
(3) Can be made more compact—and as a single enclosed and weatherproof unit—than the separate units of the coil ignition system. With the use of modern magnet alloys magnetos can be made much smaller and lighter than previously.
(4) By suitable design and construction, using modern materials, magnetos can be made to give good spark energies at low starting or cranking speeds.
(5) The high spark energies at maximum engine speeds can now be modified by the use of suitable shunts, so that the sparking plug electrodes do not burn away, as on earlier magnetos.
(6) Due to its reliability and independence of a battery, the magneto has been the standard equipment of the reciprocating-type aircraft engine over a long period of years.
(7) Absence of any low tension cables. With the coil ignition system a cable is required between the battery and coil and also the coil and contact breaker unit.

CHAPTER 5

COIL AND OTHER IGNITION SYSTEMS

HAVING dealt with the theoretical and experimental aspects of the ignition system, in the previous chapter, it is proposed to consider the more practical side of this subject.

The schematic layout of a complete ignition system for a four-cylinder engine is illustrated in Fig. 86. In this diagram it will be seen that the common or earth (ground) return method is employed, all positive connections, including that of the battery being connected to the metal of the engine block or chassis.

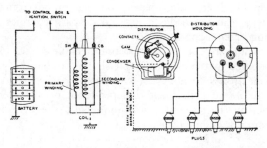

FIG. 86. Layout of complete coil ignition system for a four-cylinder engine.
Note. The distributor unit contains the contact breaker.

From the battery, when the ignition switch is closed, current flows through a primary or control resistance (see Fig. 88) to the low tension winding of the high tension coil unit, and thence to the insulated fixed contact member of the contact breaker unit. Normally, the moving or hinged arm contact is held into contact with the fixed contact, by means of the compression spring shown, but in the diagram the contacts are shown in the open position, with one lobe of the four-lobe cam pressing the hinged arm upwards. The cam is driven at one-half engine speed for a four-cycle engine, so that it produces four successive sparks every two

COIL AND OTHER IGNITION SYSTEMS 153

engine revolutions, i.e. two sparks per revolution. At the 'break' of the contacts the high voltage current in the secondary winding of the high tension (H.T.) coil is led by a high tension (H.T.) cable to the centre of a rotating arm, known as the *rotor*, which is driven at the same speed as the cam; in practice a single shaft drives both members which are arranged with their axes concentric. As the rotor rotates it passes and rubs or sparks across each of four brass contacts, in turn. Each contact is connected by an H.T. cable to a sparking plug, so that this *distributor* provides each cylinder with an ignition spark at 90° intervals, corresponding to 180° of crankshaft angle.

The condenser, shunted across the insulated and earthed contacts is shown in Fig. 86, below the four-lobe cam that actuates the contact breaker.

In connection with the high tension coil, shown schematically in Fig. 86, next to the battery the terminals are marked *SW*, for the low tension cable connection to the ignition switch, the connections being indicated by the arrows. The other terminal, marked *CB*, is for the low tension cable connection to the contact breaker.

The rotor arm member mentioned previously is not shown in Fig. 86 but the central high tension cable from the top of the high tension ignition coil leads to a central terminal with a brush connection at the centre of the distributor moulding, as shown at *R*.

Auto-transformer Method. As shown in Fig. 86 the battery

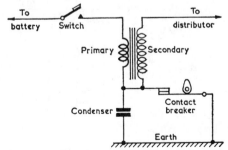

Fig. 87. The auto-transformer method of connecting the primary coil

positive or earth connection is made to the common junction of the primary and secondary coils. A better method, illustrated in Fig. 87, is to connect the live battery lead to the free end of the primary coil and the common junction of the two coils to the insulated contact of the contact breaker unit and condenser. The coil turns and therefore their voltages are then additive and the system more economical than that shown in Fig. 86.

The Ballast or Control Resistance. This is an additional resistance in series with the primary coil, as indicated at R in Fig. 88. One purpose is to limit the rate of battery discharge should the ignition be left switched on when the engine has stopped. The other and, perhaps, more important purpose is to regulate the current flowing in the primary coil. The ballast resistance is of a nickel alloy, or iron wire, having a high resistance-temperature coefficient, such that when hot, the resistance is much greater than when cold—usually two to three times as much. Thus, the resistance of the primary circuit is lower when the current is small than when it is large, so that during slow speed running when the primary current, without a ballast resistance, is relatively large, the heating effect increases the resistance and therefore reduces the current. Further, the reduction of current at high speeds is less marked. The use of a suitably designed ballast resistance enables a heavier primary current to

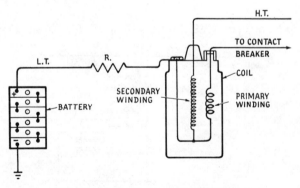

FIG. 88. The ballast or control resistance

COIL AND OTHER IGNITION SYSTEMS

be employed, and therefore greater spark energies over the operating range.

In some instances the ballast resistance is incorporated in the cable between the battery switch and primary connection on the H.T. coil.

The more recent tendency has been to dispense with the ballast resistance, due to the large variations in the cooling conditions encountered under the engine bonnet, which are apt to interfere with the action of this unit. Instead, some manufacturers employ an internally wound primary winding of higher resistance, with better cooling conditions.

Some Coil Ignition Considerations. The advantages and certain drawbacks of the coil ignition system have been mentioned in the previous chapter. One of these advantages is the ability of the coil ignition to produce a spark at the sparking plug points at engine speeds as low as 15 r.p.m., whereas the magneto requires an appreciably higher speed. Since the coil ignition system will provide more spark energy at the lower speeds than the magneto, it can produce an ignition spark under adverse conditions due to carbon deposition on the insulator and moisture, which the magneto could not provide. It is under the high speed operating conditions that coil ignition reaches its limitations.

Speed Limitation. This is due to the extremely small time intervals, at such high speeds, in which the system has to make and break the primary circuit. The limiting conditions are then those due to inductance and the tendency of the contact breaker cam to fling the moving contact arm off its surface.

Some idea of the very small intervals available for the ignition spark production may be had by considering the example of a standard design of contact breaker used with a six-cylinder engine operating at 5,000 r.p.m. Under four-cycle conditions the ignition system is required to produce a total of 15,000 sparks per min., i.e. 250 sparks per sec.

The corresponding period per spark is $\frac{1}{250}$, or ·004 sec., during which the contacts have to remain closed for a certain proportion of this interval and then open for the rest of the period. Experience and theory agree that the closed or 'make' period should, preferably, occupy about two-thirds of the total time available, and the 'break' period the other one-third. Assuming a six-lobe cam to operate the contact breaker, each lobe occupies 60°, so that the contacts are closed for $\frac{2}{3} \times 60° = 40°$.

At 5,000 r.p.m. the contacts will be closed for $\frac{2}{3} \times \frac{1}{250} =$ ·00266 sec. However, since the modern coil requires a 'make' period of ·002 sec. to produce a satisfactory spark at the plugs, if there were no other limitations the corresponding speed of 6,700 r.p.m. could be attained. With the standard design of contact breaker, the impact of the lobes of the cam on the heel of the moving arm would tend to fling the arm away from the cam, thus breaking the primary circuit for a longer period, to cause misfiring or failure to operate at this speed.

By careful re-design of the cam profile and reduction in weight of the moving arm member, together with the increase in the control spring (of the moving arm) loading, it has been shown possible to overcome this drawback and, with the aid of a specially designed coil, having the electrical performance requirements, to operate a six-cylinder engine at speeds up to 10,000 r.p.m.

Another method, referred to later, is to employ two contact breakers, so as to increase the proportion of time during which the contact points are closed. In this case the contact breakers are operated alternately by means of a single cam having one-half the number of lobes as there are cylinders, e.g. three in the case of a six-cylinder engine. This method, with six- and eight-cylinder engines has enabled the effective speed range to be increased appreciably.

With *magneto ignition*, the closed and open periods of the contacts are practically equal.

Components of the Coil Ignition System. It is proposed now to consider, separately, the components of this system. These include (1) *The High Tension* or *H.T. Coil*. (2) *The Contact Breaker*. (3) *The Distributor*. (4) *Devices for varying the Ignition Timing*, automatically, and (5) *The Sparking Plugs*.

The H.T. coil is mounted separately from the combined contact breaker-distributor unit, while the automatic ignition timing units are integral with, or attached to the contact breaker member. In addition, there are the low tension and high tension leads or cables which connect the various component terminals.

(1) The High Tension Coil. As mentioned earlier this coil is a transformer having an insulated iron core around which are wound the insulated primary and secondary windings of the two coils. Hitherto, it was usual to wind the primary coil over the core and the secondary coil over the primary coil (Fig. 89). With

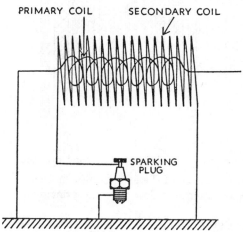

FIG. 89. H.T. coil method of winding the secondary over the primary coil

this arrangement the magnetic field is not so strong as with the later alternative method—now standard practice in this country—of winding the primary coil over the secondary coil. The mutual inductance of this latter arrangement is also higher than for the previous type.

With the primary wound over the secondary wire (Fig. 90) the length of the relatively expensive fine gauge secondary wire is reduced; also the amount of insulation between the outside of the coil and the frame, if the core is insulated from the frame. It should be mentioned that the secondary winding end is connected to the core.

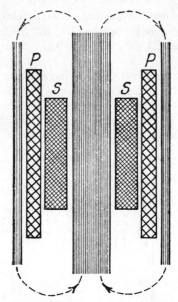

Fig. 90. Open-core coil with primary wound over the secondary coil

Another advantage of the external primary winding method is that there is better heat flow from the primary windings to the case.

When the primary coil is outside its resistance can conveniently be increased so that it is possible to dispense with the previously mentioned ballast resistance.

Two H.T. coil constructions that have been used are the *Open Core* with long air gap and *Closed Core* with a short air gap. The open core method is used with the modern external primary winding coil, as shown in Fig. 90. In regard to these alternative constructions, while both give an essential air gap in the magnetic circuit, and both can be designed to give fully satisfactory H.T. coil operation, the open core type requires more copper than the closed core type, but the latter needs more iron in the circuit.

Construction. The core is usually made of soft iron wires or laminations. When of wire the soft iron wire is of 18 to 20 S.W.G.

COIL AND OTHER IGNITION SYSTEMS 159

and if of laminations 24 to 28 S.W.G. thickness. The wires are insulated by paper or varnish and the laminations by a coating of varnish or enamel.

The primary winding usually consists of a few hundred turns of enamelled copper wire of a suitable diameter to give the coil a low resistance, namely, 0·8 to 1·5 ohms. The secondary circuit is wound with 15,000 to 20,000 or more turns of 42 to 44 S.W.G. copper wire of 2,000 to 4,000 ohms resistance. Recently with the higher voltages required at the sparking plugs the windings have increased, giving secondary coil resistances of 7,000 to 9,500 ohms.

The resistance of the primary circuit is such that in most cases, when the *engine is at rest* the *current draw* from the battery is 2 to 2·5 amps. and when the *engine is idling*, 3·5 to 5·0 amps.

In the previous dry-type H.T. coils it was the practice to impregnate the windings with liquid wax, which solidified to fill all the spaces or voids. The assembly was afterwards secured to its metal case by means of hot pitch.

It should be pointed out that if any interspaces exist, ionization may occur in the form of a slight glow discharge, although under some conditions inter-sparking may occur. A continuation of this ionic process may cause a gaseous product to be generated which in turn leaves a carbonized path so that the insulation can break down and short circuits occur in the secondary windings.

The previously described wax-filling has not proved entirely satisfactory, so that the difficulties have been overcome by using

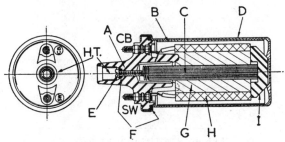

FIG. 91. The Lucas oil-filled H.T. coil. (A) moulded cover and high-tension terminal. (B) return path for magnetic flux. (C) laminated core. (D) aluminium case. (E) self-tapping screw. (F) seal. (G) secondary winding. (H) primary winding. (I) porcelain insulator

an oil-filled coil, since it is the usual practice in electrical engineering to immerse such windings in an insulating oil. Modern coils are therefore of the oil-filled kind, so that not only are any voids filled but the oil provides better heat conduction to the case.

Fig. 91 illustrates a Lucas oil-filled coil in cross-section, showing the method of construction and the various components used. A typical American coil, namely, the Delco-Remy is shown, in part section, in Fig. 92. It will be observed that in both of the illustrated coils, there is a porcelain insulator in the bottom of the case.

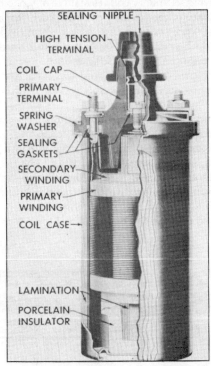

FIG. 92. The Delco-Remy H.T. coil

H.T. Coil Connections. There are three connections to the H.T. Coil, namely, two low tension and one high tension connection. In Lucas coils the low tension screwed terminals are marked 'SW' and 'CB', for connecting to the ignition switch and contact breaker casing insulated terminal, respectively. The high tension cable connection is at the central part of the top insulated cover; this cable conveys the high tension current to the distributor rotor arm, by means of a central spring-loaded carbon brush. A metal connector on the cable end is made a spring-fit in the metal sleeve of the coil cap.

(2) The Contact Breaker-Distributor Unit. The purpose of this unit is to provide the mechanism for making and breaking the low tension or primary coil circuit; to distribute the high voltage sparks to the sparking plugs at their correct sequence and to vary the spark timings according to the engine speeds. There are, thus, three principal members of this combined unit, a typical example of which is shown in Fig. 93.

The cam that operates the contact breaker make-and-break mechanism is driven at one-half engine speed for four-cycle engines and since the engine's camshaft operates at the same speed the ignition camshaft is driven off the engine's camshaft by means of a pair of helical gears. The engine drive shaft runs in a porous bronze bearing or pair of these bearings, and has the cam at the upper or outer end. The same shaft also drives the insulated high tension-fed rotor arm. The upper moulded insulator casing or distributor cover fits over the contact breaker casing, being located correctly by a dowel and notch device and secured by a pair of hinged steel clips.

Below the plate on which the contact breaker unit and the condenser are mounted is a centrifugal ignition timing device which varies the ignition sparking moments according to the engine speed, as explained later.

The housing for the distributor and cam drive shaft—not shown in Fig. 93—is provided with an adjustment such that the initial setting of the ignition timing, when the engine is at rest, can be adjusted and then locked in position.

The Contact Breaker. The object of this unit is to make and break the primary circuit of the H.T. coil so as to produce one

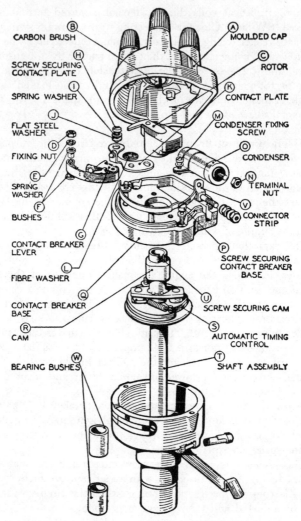

FIG. 93. Components of contact-breaker and distributor unit (Austin)

spark per cylinder, once every two engine revolutions. As explained earlier sufficient energy is stored in the magnetic circuit when the contacts are closed, to provide the spark energy when the contacts are opened. The interval between successive sparks becomes extremely small as the engine speed increases, until at a certain critical interval, namely, ·004 sec., corresponding to an engine speed of 5,000 r.p.m., in the case of a six-cylinder engine, the contact moving arm or lever may no longer remain in contact with the cam face but will be thrown off, by centrifugal action overcoming the tension of the contact lever spring, in the case of a standard contact breaker unit. It is found by experiment that a maximum separation of the contacts of about ·015 in. gives the best results in practice. With a parabolic operating cam the contact lever on impact of the cam is accelerated, reaching its maximum velocity and then returning to zero by the time the contacts are fully opened, i.e. at ·015 in.; the lever under the spring action accelerates and returns to zero speed as the contacts close. Above the critical speed, the contact lever is thrown clear of the cam and the contacts gap is thereby increased.

In order to extend the initial speed of the contact lever, before flinging action occurs, the lever itself can be made lighter, so as to reduce the inertia effect, and the spring can be made stronger. Further, the characteristics of the H.T. coil must be chosen so that the sparking voltages occur satisfactorily as the available interval between the individual sparks is decreased—due to increased engine speed.

Fig. 94 shows the results of an investigation into contact breaker and H.T. coil operation due to E. A. Watson, in which the relation between distributor, or cam speed and the times to open and close is shown graphically. It will be seen that there is a definite limit of about 2 milliseconds (·002 sec.), so that even if the H.T. coil could produce sparks at 1 millisecond intervals, there would be a very small gain in the operating speed.

It can be shown that at 2 milliseconds sparking intervals, the speed can only be raised from 5,000 to 6,700 r.p.m. by increasing the spring tension by 77 per cent, or reducing the contact arm inertia in the same ratio.

It is possible to provide the H.T. coil with an external

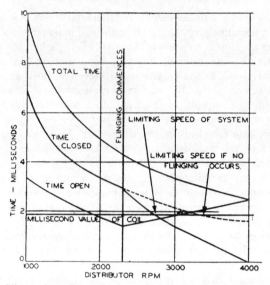

Fig. 94. Illustrating effects of contact breaker flinging at limiting speeds

resistance, to give an interval of approximately 1·5 millisecond (·0015 sec.) which with the contact points closed angle of 40° gives a time per camshaft revolution of $\frac{360}{40} \times 1\cdot5$, or 13·5 milliseconds and an engine speed of 8,900 r.p.m. It will, however, be necessary to increase the tension of the contact lever control spring by 3·15 : 1. In this connection it should be mentioned that the spring loading increases as the square of the cam speed.

Contact Breaker Design. The modern unit (Fig. 95) employs a resin-loaded fabric or glass-filled nylon contact breaker lever, instead of the earlier red fibre material, which was subject to hygroscopic effects which sometimes caused sticking of the lever on its pivot pin. The moving contact, of tungsten, is welded to a steel strip member, firmly attached to the lever material. The control spring is now made of stainless steel and is attached to the low tension terminal fitting, another metal arm from which is connected to the insulated side of the condenser. It is current

COIL AND OTHER IGNITION SYSTEMS

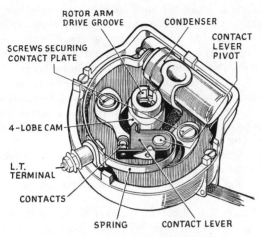

Fig. 95. Typical contact breaker unit, with distributor cover removed (Lucas)

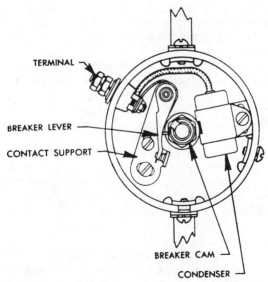

Fig. 96. Typical American contact breaker unit (Delco-Remy)

British practice to make the steel strip to which the contact is welded, flexible to a limited extent, so that a small amount of *sliding action* occurs between the mating faces of the contacts when they meet; this 'wiping' action extends the life of the contacts appreciably, before they need trimming again.

Fig. 96 illustrates an American unit, in which the contact breaker lever is of channel-section thin steel, and has the tungsten contact welded to its free end. The steel lever has a resin-bonded heel for contacting the cam securely attached to it; this design gives the wiping action of the British unit.

The operating cam, which has as many lobes as the engine has cylinders, in the half-engine speed cam model, is designed as a cam to give uniform acceleration and deceleration.

The Condenser. This important member of the ignition system the functions of which have previously been described, was originally an external unit but in modern practice is mounted within the contact breaker casing. It is usually of cylindrical shape with a metal casing, forming the earthed side and an insulated cable or bar connecting the condenser 'live' terminal to the insulated low tension terminal on the side of the contact breaker casing.

In principle, the condenser consists of two metal plates between which is an insulator or dielectric material. In practice the condenser consists of two sets of tinfoil or aluminium strips, usually wound spirally, one around the other, and separated by a dielectric such as mica or strips of paper, oiled, varnished or waxed.

Previously, coil ignition condensers were made of rolled rice or tissue paper sandwiched between strips of tinfoil which overhung on opposite ends, to give a tubular construction. Terminals were soldered to each tinfoil end. Low resistance was achieved by feeding the current into the long edge of the strip. The condenser was vacuum-dried, wax-impregnated and sealed into a metal casing.

The modern condenser is of the self-sealing kind having a wound strip of paper upon which a thin metal film has been vacuum-deposited, so as to give a continuous metal surface, for each element. The advantage of this method is that it not only gives a very good condenser unit, but is self-sealing, i.e. if a hole is made in the paper the metal film around the hole is evaporated.

The alternative method that has been used on American ignition equipment is to use impregnated paper and aluminium foils, wound around one another.

The *insulation resistance* of the condenser should be high, namely, at least 10 megohms between the plates and should be capable of withstanding 1,000 volts. The insulation resistance can, however, fall to 1–2 megohms and the condenser will perform satisfactorily.

The capacitance of ignition condensers, as stated earlier, usually lies between ·15 and ·35 microfarads, a common value being 0·20 to 0·25 mfd.

Cam Angle. From previous considerations of ignition sparking intervals it will be understood that since these depend upon the number of lobes on the contact breaker actuating cam and also on the engine (or camshaft) speed, it is important that the cam should be designed so as to give the correct ratio of 'closed' to 'open' contacts interval. Especially must the closed contacts interval be as long as practical, to allow the magnetic field to build up to its maximum strength. In this connection it is usual to define the angle between which the contacts remain closed, as the *cam* or *dwell* angle. This angle will be less than the angle between the cam lobes due, of course, to the opening period of the cam.

For a six-cylinder engine (Fig. 97) the cam angle is usually from $32°$ to $37°$, whereas the actual angle between the centres of the cam lobes is $\frac{360}{6}$, or $60°$.

For vee-eight engines the cam angle is about $26°$ to $30°$ and the lobe angle, $\frac{360}{8}$ or $45°$. The cam angle depends also upon the maximum separation distance of the contacts, so that it is important to adjust the contacts to the minimum gaps recommended by the manufacturers. Thus, whereas in

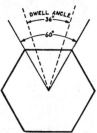

FIG. 97. The cam, or dwell angle, for six cylinder engine

some cases gaps of ·018 to ·022 in. are used, the cam angle can be increased appreciably by using the usual minimum value of ·012 in.; an increase in the mean value to 42° from about 35° is possible in this example.

In some ignition units provision is made for quick adjustment of the gap, to provide the correct cam angle, a window being fitted to the distributor unit, to show the correct value, when the screw adjustment provided is used.

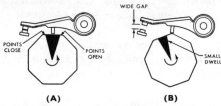

FIG. 98. Showing effect of cam angle on contact breaker point gap

Fig. 98 (A) illustrates the normal dwell angle and contact breaker gap, in the case of the Mercury eight-cylinder vee-type engine. In this case the dwell angle is 26° to $28\frac{1}{2}°$.

The result of too small a dwell angle as shown at (B) in Fig. 98 is to give *too wide* a contact gap. If the dwell angle is too great, then the contact gap will be *too small*.

A special 'Tach-Dwell' meter is used to measure the dwell angle of the ignition unit.

(3) The Distributor. As outlined earlier in this chapter, the purpose of this unit is to distribute the ignition sparks, at exactly

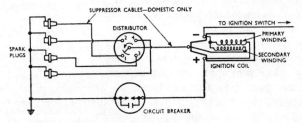

FIG. 99. Schematic layout of the distributor system (Vauxhall)

COIL AND OTHER IGNITION SYSTEMS 169

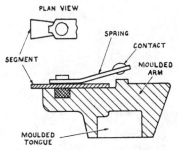

Fig. 100. The Rotor Arm unit

their correct firing intervals, to the individual cylinders, of multi-cylinder engines. The general principle of the distributor system is shown schematically in Fig. 99. for a four-cylinder engine system. The contact or circuit breaker is shown below in this diagram, with the H.T. coil on the right. The H.T. cable from this coil is connected, *via* a spring contact to the central part of the H.T. rotor arm which is driven from the same shaft as the contact breaker cam (Fig. 100). Previously, the outer end of the rotor metal member made contact with each brass segment inside the distributor cap, as it rotated across them, but now a small gap is arranged between the rotor arm and the segments so that the H.T. current 'jumps' across the gap instead of by the brushing contact method. This method results in less wear of the metal members.

Since each brass segment is connected to the conductor of a cable leading to one of the sparking plugs, an ignition spark occurs whenever the rotor is opposite a metal segment of the distributor.

The distributor unit forms the cap over the contact breaker, being held in its correct position by a locating dowel or tongue which engages with a slot in the fixed metal casing of the contact breaker assembly and held in position by a pair of hinged spring clips one of which is shown in Fig. 93. In this Lucas design the H.T. cable from the coil is electrically connected to a spring-loaded carbon brush, which makes a light contact at the rotational centre of the metal member of the insulated rotor arm, so as to convey the H.T. current to the rotor at all times.

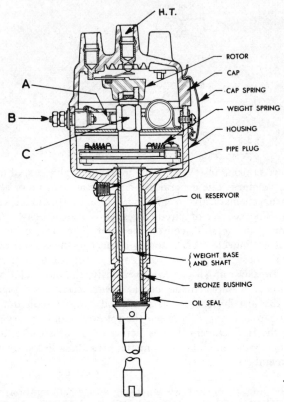

Fig. 101. Delco-Remy contact breaker and distributor assembly

Fig. 101 shows a Delco-Remy (American) unit, comprising the engine half-speed drive shaft and its metal housing, the centrifugal governor weight spring—to which reference is made later—the contact breaker cam C and contact breaker assembly A, the low tension terminal B and the internally-ribbed moulded distributor cap. The H.T. current from the coil is led to the central terminal of the cap and is electrically connected to a spring arm attached to the top of the rotor arm, whence the H.T. current is conveyed, at all times, to the metal part of the rotor arm.

COIL AND OTHER IGNITION SYSTEMS 171

The vertical shaft runs in porous bronze bearings, supplied with oil from a tubular reservoir.

The rotor arm is usually moulded from a phenol resin or similar high insulation product and it is made with a moulded tongue inside the hole, which fits over a recess cut at the end of the distributor or cam shaft, so as to afford a positive drive for the rotor arm and at the same time enable the arm to be detached from the shaft when necessary; it is made a tight sliding fit for this purpose. For heavy duty rotor arms the segment outer end is coated with tungsten, in order to withstand the arcing effects of the 'jump spark' over long periods of operation.

Notes on Distributors. The distributor cap must be designed to obviate certain faults liable to occur under modern high voltage and engine speed conditions.

One such trouble sometimes met with in previous designs was a tendency for the H.T. current to *track across* from the central cable connection to one of the plug cable segments or between adjacent segments. This effect is shown by fine tracks on the insulation surface and it results in engine misfiring at the higher speeds.

Ventilation of the space enclosed within the cap is important, since one result of the jump spark method is to create a conducting nitrous powder due to ionization effects. Ventilation holes should be provided to allow the escape of the ionization products.

Erosion of the metal at the end of the rotor sector and the fixed cap metal sectors to which the plug cables are attached is another trouble that is liable to occur. By making the high voltage supply to the rotor metal *positive* the sector end will burn away more rapidly than if the supply is negative, so that with the positive earthing of the battery the rotor arm metal will have a longer life.

(4) Ignition Timing. In order to develop the full power output of a petrol engine at any given speed it is necessary for the ignition spark to occur at the correct moment during the latter part of the compression stroke. Unless this is arranged the engine will lose power, e.g. if the spark occurs earlier or later.

If the maximum pressure and temperature due to the ignition of the compressed air-petrol charge were to be developed instantaneously it would only be necessary to 'time' the spark to occur when the ascending piston had just reached its top dead centre (T.D.C.). However, since there is a definite time lag between the moment of occurrence of the spark and the attainment of maximum pressure, this must be allowed for by arranging the ignition spark to occur some time before the piston reaches the end of its compression stroke. Depending upon the engine design, compression ratio, speed and other minor factors the equivalent crank angle may vary between about 20° and 40° for the lag period, i.e. the spark must take place some 20° to 40° of crank angle before the piston reaches its T.D.C., on the compression stroke.

Using the Indicator Diagram. The most accurate way to ascertain the correct ignition timing would be to study the pressure-volume (or indicator) diagrams of an engine running at constant speed, throttle opening and mixture strength, and then measure the area of the indicator diagram. That giving the maximum area will correspond to the greatest power output and, therefore, the correct ignition timing.

In this connection reference should be made to the three indicator diagrams given previously, in Fig. 76, showing the effects of correct ignition timing, advanced and retarded timing. As this subject is discussed in Chapter 4 it is unnecessary to dwell upon the subject further.

In regard to the practical aspects of ignition timing it may be stated that apart from a loss of power when the ignition is *over-advanced*, there will be a tendency to 'rough' operation with probable *detonation* effects, due to the more rapid rates of pressure rise and the appreciably higher pressures attained.

When the ignition is *retarded* in relation to its correct position, not only will the power output be reduced but, due to late combustion, the engine will tend to *overheat* and if there is too much ignition retard, the combustion chamber may become *too hot* and eventually cause pre-ignition, i.e. premature ignition of the compressed charge before the ignition spark occurs. It may be added that too much ignition advance can cause overheating.

COIL AND OTHER IGNITION SYSTEMS

Effect of Engine Speed. From both theoretical and practical considerations it has been shown that if the ignition timing is set correctly for a relatively low engine speed then, as the engine is accelerated it is necessary to advance the ignition timing, progressively, in order to develop maximum power.

In the earlier car and motor cycle engines an ignition control or advance lever was provided, to enable the driver to advance or retard the ignition timing according to road speed—and also to some extent, load—conditions.

In modern ignition units the ignition is advanced, automatically, as the engine speed is increased from fast idling, by a centrifugal device which is interposed between the driving and driven shaft of the contact breaker section. When the engine speed is increased hinged weights, having control springs, are moved outwards by centrifugal action so as to change the angular relation of the driving and driven shafts, i.e. to increase the period between the ignition spark and maximum pressure, or the T.D.C. position of the piston.

It is possible to obtain the desired shape of the ignition advance-engine speed curve by varying the strengths of the control springs.

Assuming that the initial ignition timing with the engine running at idling speed is correct, the centrifugal mechanism should advance the timing as the engine speed increases, approximately in the manner depicted in Fig. 102, in which there is a curved portion CD of the graph, corresponding to a more rapid

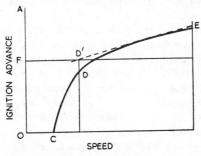

Fig. 102. Illustrating effect of engine speed on ignition advance

increase of the timing from its retarded condition at the idling speed OC. The portion DE of the graph is practically straight, for full-throttle conditions; in this connection the centrifugal mechanism can readily be arranged to give the characteristic line D'E.

In the case of a typical centrifugal advance mechanism for a modern car engine with a compression ratio of 9·0 : 1 and maximum engine speed of 4,000 r.p.m. the ignition advance graph

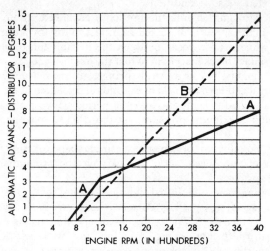

FIG. 103. Engine speed, spark advance graphs for (A) Cadillac Vee-eight engine. (B) Vauxhall four-cylinder engine

is as shown in Fig. 103 (A). In this case, the mechanism commences to operate from the idling speed of about 700 r.p.m., giving a 3° advance, i.e., 6° in the crankshaft at 1,200 r.p.m., after which the advance increases linearly to 8° at 4,000 r.p.m. It may be mentioned that the maximum crankshaft advance angles, according to the design of engine and its operating conditions can range up to 30°–40°. The advance angles at different speeds are checked with a special apparatus both for increasing and decreasing speeds. Fig. 103 (B) is a corresponding graph for a 92 cu. in. four-cylinder engine.

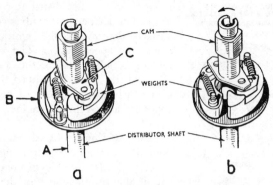

FIG. 104. Centrifugal ignition advance mechanism. (a) Static position. (b) Fully-advanced position

Ignition Advance Mechanism. The principle of one type of centrifugal mechanism, namely the Lucas, is shown in Fig. 104 in which it will be seen that the engine-driven distributor Shaft A drives the base plate B on which are mounted two centrifugal weights which are hinged at one end and restrained by the springs shown. The two weights have pins C which engage with holes in the upper flange member D which is integral with the cam that operates the contact breaker arm.

As shown on the left hand the mechanism is in its static position. When the engine is running and its speed is increased the weights are moved outward at their free ends and in doing so their pins C turn the flange member—and with it the cam—in an anti-clockwise direction as shown by the arrow in the right hand illustration. This causes the contact breaker points to open earlier than for the slow speed position and therefore advances the ignition timing. As the engine speed decreases the forces on the weights diminish and the control springs assist, to retard the timing.

The centrifugal mechanism of the Delco-Remy automatic advance unit is shown in both the static and maximum speed positions in Fig. 105. The mechanism consists of an *advance cam* which is integral with the cam that operates the contact breaker, a pair of advance weights and springs and a weight base which is

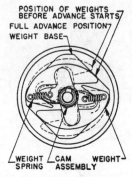

FIG. 105. Delco-Remy ignition advance mechanism

integral with the distributor drive shaft. The shape of the advance cam surface determines the ignition-advance and speed curve. The toggles on the weights actuate the advance cam and therefore the contact breaker cam which, in this case is of hexagon shape for a six-cylinder engine.

Effect of Engine Load. Since it is possible to operate a petrol engine at different loads, for any given engine speed, the combustion and therefore the ignition timing characteristics should be taken into account. In this connection it is necessary to introduce an additional ignition timing control for variations of throttle opening, or engine load or torque.

This is effected in modern ignition units by utilizing the variations of the vacuum in the engine inlet manifold, since it is known that the inlet manifold vacuum varies very nearly inversely as the engine load. Thus, at low loads or throttle opening the vacuum is at about its maximum value, namely, 18–20 inches of mercury (Hg.), while at high loads the value falls to minimum values of a few inches of mercury.

Fig. 106 shows the relation between engine power, i.e. throttle opening, and ignition advance, for a given engine speed, denoted by the value OA. The four curves refer to the throttle openings marked alongside. At full throttle, the centrifugal mechanism advances the ignition to the point A the maximum power being shown by the ordinate AP_1.

At part throttle openings the operating points will still lie on the line AP_1 since the centrifugal advance is constant. The peaks of the corresponding power curves do not, however, lie on this line but on the curve P_1, P_2, P_3, P_4. It will therefore be seen that it is necessary to provide an additional advance when operating at part throttle, to enable the engine to develop its full power. Usually, the maximum advance obtained from the vacuum or load control is from 14° to 16°, on the distributor, but values as low as

COIL AND OTHER IGNITION SYSTEMS 177

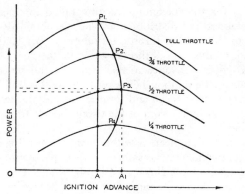

FIG. 106. Relationship between engine power, throttle opening and ignition advance (Lucas)

9° and as high as 24° have been used in certain automobile engines.

The load advance graph for the Cadillac engine, corresponding to the centrifugal advance graph, shown by the AA graph in Fig. 103, is given in Fig. 107. It will be seen that the load control

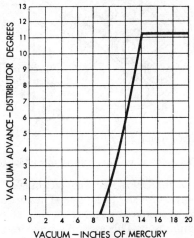

FIG. 107. Vacuum advance graph (Cadillac)

commences for the maximum vacuum of 20 in. mercury, for which just over 11° distributor advance is obtained; this corresponds to the smaller throttle openings or part loads. At full load, corresponding to full throttle and 9 in. mercury, there is no additional ignition advance to that given by the centrifugal, or speed control, the ignition being retarded as the load increases. At any given engine speed and throttle opening the *total ignition advance* is given by the sum of the two ordinates from the speed and load curves.

The method of application of the vacuum control is to employ a diaphragm mounted in a casing, one side of the diaphragm being connected to a simple linkage which is arranged to rotate the distributor housing in its mounting. The other, or vacuum side of the diaphragm chamber is connected by means of a metal tube to the inlet manifold. The diaphragm is spring-loaded on the vacuum side, so that as the degree of vacuum is reduced the spring returns the diaphragm to the full-load or 'zero-advance' position, thus leaving only the speed control in operation.

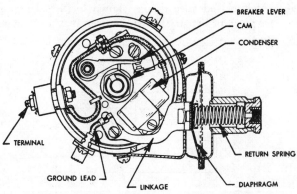

FIG. 108. Delco-Remy vacuum control to contact breaker unit

Fig. 108 shows the Delco-Remy contact breaker assembly with the vacuum control unit and its linkage. The vacuum supply tube is connected to the screwed member on the extreme right hand side of the vacuum unit. This assembly provides for a maximum centrifugal advance of 30° (crankshaft) at 2,700 r.p.m. for the

COIL AND OTHER IGNITION SYSTEMS

American Motor Corporation six-cylinder engine and a maximum vacuum or load control advance of 12° for 16 in. of mercury.

Timing the Complete Ignition Unit. While the speed and load control devices take care of the ignition timing once the engine is running it is, of course, necessary to determine the static setting of the complete ignition unit, in relation to engine crankshaft.

According to the engine design and the grade of fuel used it is usual to set one of the cylinder pistons—usually No. 1—on its top dead centre and then rotate the engine crankshaft forwards, to the ignition timing angle specified by the engine manufacturers for the grade of fuel used.

When the crankshaft is in this ignition timing position, the contact breaker points should just be separating, i.e. the primary circuit is 'breaking' and the ignition spark occurring. It is assumed that the distributor shaft driven cam is rotating in the correct direction and that No. 1 piston is nearing the end of its compression stroke.

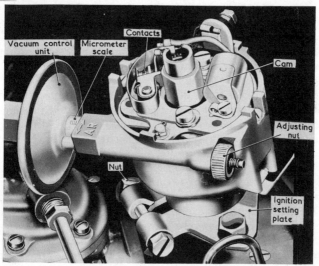

FIG. 109. Vacuum control and micrometer adjustment screw (Ford)

Since the distributor shaft is gear-driven from the engine cam shaft, it is necessary to rotate the complete distributor unit slightly about the distributor shaft in order to vary the initial timing setting when the ignition unit is assembled on the engine. The Lucas units are provided with a clamping plate, held by a screw and nut so that on releasing the nut the complete unit can be moved about the shaft axis, in order to advance or retard the ignition setting.

Once the ignition timing has been set, it is possible to make small angular adjustments by means of a micrometer screw (Fig. 109) which is provided with a graduated scale and index, on the vacuum control unit connection. The arrows and letters A and R denote the direction of rotation of the adjusting screw nut to advance or retard the ignition. The minor adjustment is usually employed for altering the timing when the grade of fuel used is changed. Usually, if the initial ignition timing for non-premium fuels is, say, $4°$ before T.D.C. then on premium fuels, it will be from $6°$ to $8°$ B.T.D.C.

Limitations of the Coil Ignition System. More recently the tendency in automobile engine design has been in two principal directions, namely, (1) The use of increasing compression ratios, e.g. from about $8·5 : 1$ to $10·5 : 1$. (2) Adoption of higher maximum engine speeds, e.g. 5,000 to 6,000 r.p.m. in production cars and up to 12,000 r.p.m. for racing cars and motor cycles.

The reasons for these developments are concerned not only with the production of greater power outputs per unit cylinder capacity, i.e. the H.P. per litre, but also with the demand for minimum fuel consumptions; in this connection it is well known that as the engine compression is increased, so is the thermal efficiency. Since the fuel consumption varies inversely as the thermal efficiency it follows that with compression increase, fuel consumption diminishes. Thus, the fuel consumptions of higher compression engines as expressed in terms of lb. of fuel consumed per B.H.P. per hour are relatively lower.

The battery and coil ignition system has certain limitations at the higher engine speeds and compressions which can be sum marized, briefly, as follows:

COIL AND OTHER IGNITION SYSTEMS 181

(1) Inability of the contact breaker to operate at these higher speeds, due to mechanical troubles. Thus, on this and other counts, present systems limit the equivalent speeds to values corresponding to about 400 sparks per sec., i.e. a time per spark of $\frac{1}{400}$ sec., or ·0025 sec.
(2) Plug fouling, due to the effects of the lead salts used in petrols to increase their suitability, i.e. octane ratings, for the higher compression ratios so that detonation effects cannot occur.
(3) Necessity to employ ever increasing high voltages to produce ignition sparks at the higher engine speeds and compressions, e.g. ignition system voltages of 20,000 and above.
(4) Contact breaker points pitting, or burning, due to the high currents used.
(5) Inaccuracy of ignition timing at higher speeds due to backlash and torsional oscillations in the drive mechanism.
(6) Limitation due to reduced cam dwell times, associated with cam design and contact breaker operation at higher speeds.

The present ignition systems, described earlier, have been steadily improved by the use of better ignition coil characteristics, cam design, improved cam angle and double contact breakers, so that something approaching a limit has been reached in their performances, therefore, the attention of ignition system manufacturers has been directed to the possibilities of alternative systems for higher performance engines. Some brief particulars of the more promising of these will now be considered.

Low Voltage, Surface Discharge Plug Systems. A lower voltage ignition system which was developed to overcome the difficulties with high voltage systems at high altitudes in the case of aircraft engines, was developed by a Dutch engineer, W. B. Smitzs, after the last World War.

The system which operates with a secondary voltage of less than 3,000 volts, uses a condenser in the secondary circuit and a rectifier (to produce direct current for charging the condenser) between the coil and condenser. The timing of the ignition discharge is done by the distributor, instead of by a contact-breaker mechanism. The current obtained from the charged condenser is taken by the rotor to fixed contacts in the distributor housing, in the usual manner. The current flows only when the contact occurs between the rotor and the fixed contacts, to which the sparking plug cables are connected.

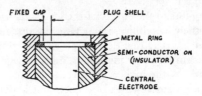

FIG. 110. Spark gap end of surface discharge plug

Instead of the usual single point sparking plug, another kind, known as the *surface discharge* plug (Fig. 110) is employed. This plug has a fixed sparking gap consisting of a central cylindrical electrode on a concentric metal ring which forms part of the plug shell. The usual plug insulator is replaced by a semi-conducting ceramic at the insulated tip. This ceramic has a negative temperature coefficient, i.e. the resistance of the material falls with increase in temperature.

Once the ignition system high energy discharge occurs, the space between the exposed ends of the inner and outer electrodes becomes ionized and the 'arc' between these electrodes across the semi-conducting surface of the insulator is established. Since this 'arc' can occur anywhere between the concentric electrodes the wear is greatly reduced, relatively to that of the ordinary sparking plug. During the spark period the current is relatively high but, as stated the voltage is appreciably lower than for the ordinary ignition system. The main advantages of the surface discharge plug, when actuated by a suitable high energy system are as follows:

(1) Lower operating voltages after initiation of the surface discharge current.
(2) The operation of the plug is unaffected by soot or carbon formation, while moisture on the outside insulator has no detrimental effect. In practice the effect of a layer of carbon

between the electrodes can actually improve the electrical performance.
(3) Suitability—with the appropriate electrical system—for the demands of high performance engines.
(4) With the fixed and relatively long electrode gap, plug point wear is very considerably lower.
(5) The plug runs 'cold' and therefore does not have to attain the self-cleaning temperature of the orthodox type of sparking plug; the plug can therefore be used on a wide range of petrol engines.
(6) It is particularly suitable for two-cycle engine ignition purposes, since it eliminates the special plug troubles often experienced with these engines.

Gas Turbine Fuel Igniters. Surface discharge or 'high energy' plugs are widely used for igniting the fuel spray in the combustion chambers of gas turbines. The plug used is made from a tube of nimonic alloy fitted concentrically around a metal rod. These form the electrodes, being separated by an insulator of Sintox which has a semi-conductor material coating at the firing end. The ignition system consists basically of a condenser of about 6 mfd. capacity which is charged, by a kind of trembler coil, to about two kilovolts and storing energy of about 12 joules. The condenser is then discharged across the semi-conductor surface of the plug, thus providing an ignition spark which cannot be quenched by oil, fuel or water.

The H.E. plug is also used to ignite a small fuel-fed torch igniter for gas turbine ignition; in this case the jet of fuel when ignited produces a flaming torch which quickly ignites the main body of fuel supplied to the combustion chamber burners. (Vide Fig. 145).

More recently, as described later, transistorized ignition systems have enabled standard-type distributors and contact breaker units to be used with surface discharge plugs.

Transistorized Ignition Systems. In the quest for ignition systems to operate high speed engines with compression ratios above about 10 : 1, the limitations of the ordinary coil and magneto ignition systems have led to the investigation of transistor systems that would overcome these limitations. These systems were found to offer decided advantages in handling the increased currents required by the higher performance engine ignition demands.

The transistor, which may here be regarded as a type of crystal which, like the radio valve, has the power of amplifying currents, is now widely used by electrical engineers. Due to its much smaller size than the valve, its robust quality, low power consumption and small heat dissipation, it is particularly adaptable to a wide field of electronic purposes.

In its present application, the transistor gives very good amplification with an extremely clean cut-off, and therefore is particularly suited to efficient switching operations. It can be used in the primary circuit of an ignition coil, it can act as a make-and-break, to produce the maximum inductive effect in the secondary circuit. Moreover, only a very small current, namely, about 4 per cent of the usual load, is necessary for its control, so that the contact breaker which is still necessary for timing and circuit make and break purposes has only to deal with a much smaller current; therefore, burning away of the contacts is greatly reduced.

In connection with the practical application of transistors to ignition requirements some important developments were made earlier by the Electric Autolite Company[*] of America and a satisfactory system applicable to automobile engines was produced.

Of the two alternative transistors available, namely, silicon and germanium, the latter was adopted. This transistor will handle currents of 30 amps., but will not tolerate the reverse current voltages of the conventional ignition coil—which may be as high as 400 volts—so that a special coil or H.T. transformer was employed; this was termed a 'slow' coil.

Figs. 112 and 113 show the transistorized voltage ignition system which gives a relatively flat voltage output graph throughout the engine speed range, with reduced primary current which gives practically unlimited life and reliability to the contact breaker points, with absence of plug fouling.

[*] G. E. Spaulding, Jr., *Soc. Autom. Engrs. Journal*, Vol. 68, p. 610, 1960.

Referring to Fig. 111 the principle of operation is based upon the transistor's property of amplifying current. When the contacts C close a small current flowing through the emitter base circuit (Fig. 111) will switch on the transistor thus allowing a larger current to flow through the 'emitter-collector' unit.

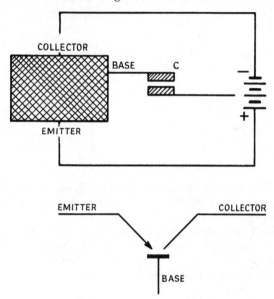

FIG. 111. Schematic diagram, to illustrate transistorized system

In more detail, the operation of the system (Fig. 112) is as follows: The diode D_1 provides a means of reverse biasing the emitter to the base junction when the contact breaker points are open. This ensures satisfactory transistor switching, even at high temperatures. The resistor R_1 allows a small current to flow through D_1 continuously, so that there is a voltage drop of 0·5–0·75 across D_1. As the transistor base is connected to the positive side of the diode through R_2 and the emitter is directly connected to the negative side of the diode, when the contacts are open, the base is at a potential of 0·5–0·75 volts positive, with respect to the emitter; this ensures the transistor switching or

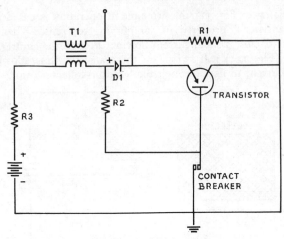

FIG. 112. Transistorized high voltage system

cut-off. This action is enhanced by keeping R_2 as small as possible, although R_2 should not be too low, otherwise it would allow too much current to flow through the contacts on the 'make' operation. During the 'dwell' period when the contacts are closed the transistor base is connected direct to the collector (Fig. 111) thus allowing the maximum current to flow. The initial current flow is limited by the induction of the H.T. transformer T_1, but finally by the resistance R_3.

It will be seen that the usual condenser, that is shunted across the contacts, in the orthodox coil ignition circuit, is no longer used in the present system.

Fig. 113 shows a comparison between the conventional ignition (A) and the transistorized high voltage ignition circuit (B) as applied to the ordinary type sparking plug; the latter system is now developed for the surface discharge or surface-gap system.

The comparative available voltages of the two systems are shown by the graphs in Fig. 114 from which the almost constant and higher voltage production of the transistorized system is shown to advantage over the lower and falling voltage production of the conventional coil ignition system.

COIL AND OTHER IGNITION SYSTEMS

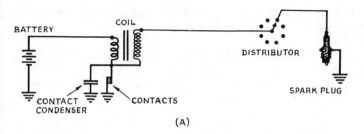

(A)

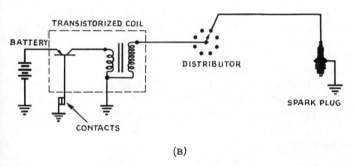

(B)

FIG. 113. Comparison between conventional ignition system (A) and transistorized high voltage system (B)

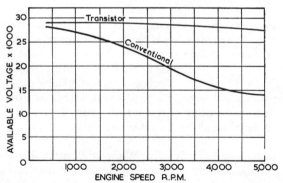

FIG. 114. Performance curves for conventional and transistorized ignition systems

The advantages of the transistorized ignition system described may be summarized, as follows:

(1) Elimination of burning of the contact breaker points over long periods.
(2) Production of a higher voltage which remains practically constant over the engine speed range, giving much better top speed performance in high-speed engines.
(3) Possibility of operating at still higher voltages should future engine development require such voltages.

The results of tests made with two similar car engines fitted, respectively, with conventional and transistorized ignition systems showed that after 1,500 hours the contacts of the former engine system were badly eroded and almost useless for further operation. The contacts on the transistorized system were not eroded but only slightly discoloured; it was estimated that the contacts of the transistorized system would have a useful operating life of 100,000 miles before total replacement was necessary.

The chief drawbacks of the system described are briefly as follows:

(1) The maximum engine speeds are still governed by the limitations of the contact breaker mechanism, e.g. flinging of the moving arm at higher speeds.
(2) The pack or unit containing the H.T. transformer, transistor, resistances, etc., is comparatively expensive relatively to the conventional ignition units.
(3) The pack is not designed to operate at ambient temperatures, e.g. those of the engine bonnet space, exceeding 350° F.
(4) The system is designed for negative earthing as compared with the positive earthing of the system which is standardized in Great Britain and certain other countries.

Transistor Ignition System Without Contact Breaker. It has been pointed out that the contact breaker of the conventional ignition system definitely limits the maximum number of sparks per second to about 400, and therefore the corresponding engine

COIL AND OTHER IGNITION SYSTEMS

speed; in the case of the magneto the limit is rather higher. To avoid this limitation, in another type of transistorized ignition, the contact breaker unit is dispensed with and in its place an electromagnetic triggering device, associated with the engine flywheel is employed. This impulse generator which is independent of centrifugal effects and torsional vibrations gives very accurate timing and a much higher spark rate. Thus in the Lucas system the spark rate is 1,000 per sec., which is equivalent to an eight-cylinder engine speed of 15,000 r.p.m. As with the previously described transistorized system the voltage is sensibly constant over the entire speed range.

In the Lucas system there is an electromagnetic pick-up associated with pole pieces fixed to the engine flywheel or a light wheel at the timing case end of the engine, a trigger amplifier, a spark generator unit and a high voltage distributor. As the engine operates, a voltage impulse is applied to the trigger amplifier—which may be regarded as a normally-closed 'switch' allowing current to flow through the primary circuit of the transformer. The effect of the voltage impulse is to open this 'switch' so that the current flow stops.

The energy released by the collapse of the current induces a voltage in the trigger transformer secondary winding which causes current to flow in the base circuit of the spark generator unit. An associated transistor thereby becomes conducting so that current flows in the primary winding of a high voltage transformer. This initiates a regenerative oscillation, causing a very rapid increase in the primary current and this, in turn, gives rise to an induced voltage in the transformer secondary circuit of over 20,000 volts. This is fed to the rotor arm of the distributor which directs the high voltage current to the sparking plugs in turn. Regeneration stops when the transformer is saturated and the transistor again becomes non-conducting.

The complete cycle time for regeneration is less than ·0002 sec. With the stopping of the voltage pulse at the pick-up, conduction commences again in the trigger amplifier circuit, ready for the cycle repetition at the next pick-up impulse.

The layout of the Lucas electronic ignition system is shown schematically in Fig. 115, for a four-cylinder engine.

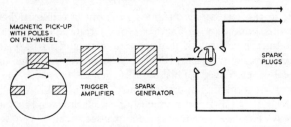

FIG. 115. The Lucas electronic ignition system for a four-cylinder engine

The running current is proportional to the engine speed, rising from about 0·25 amp. at 50 sparks per sec. to 2·5 amps. at 500 sparks per sec.; this is the inverse of the conventional coil ignition system. Thus, at low engine speeds very little current will be taken from the battery while at medium and higher speeds the cut-out will have operated and the dynamo will be generating all the electrical output for ignition and other automobile requirements.

The various components of the Lucas system—which was designed primarily for high performance engines, using compression ratios of 10 : 1 and above and operating at higher engine speeds—are shown in Fig. 116; it will be apparent that the complete equipment is appreciably more costly than the two-unit coil ignition one.

Delco-Remy Method. Instead of using a flywheel-type triggering device the Delco-Remy system employs a distributor-shaft-driven four-lobe iron timer which rotates between the poles of a ring-type permanent magnet. Thus, the magnetic field alternately builds up and collapses, so that a voltage pulse is induced in the pick-up coil. Each pulse is conducted to the transistor control unit, where it is applied to the triggering transistor, causing it to turn off the switching transistor. This halts the current flow through the ignition coil primary winding, causing the H.T. winding to produce the ignition spark at the sparking plug. No contacts or condensers are used in this system.

The Piezo-electric Ignition Method. Another more recent alternative to the usual coil ignition system makes use of the

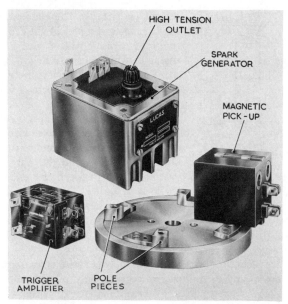

Fig. 116. Components of the Lucas electronic ignition system

piezo-electric property of certain crystals which when subjected to compression generate an electric charge. When such crystals are placed in an electric field they experience changes in size. These properties are utilized fairly widely in pick-up units used on gramophones, microphones, etc.

In the application of the compressibility electrical property to self-contained ignition units for petrol engines—at present of the single or twin-cylinder two- and four-cycle types—the earlier crystals, e.g. quartz, lithium sulphate, Rochelle salt, have been replaced by one of the piezo-electric ceramics, e.g. barium titanate or lead zirconate-titanate; the latter has proved more suitable in regard to higher temperatures, lower dielectric losses and reduced depolarization rate.

The principle of the ignition unit—which can be made very compact—is that of the piezo-electric crystal or element and a simple lever mechanism for applying the compressive load to the

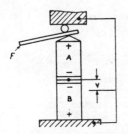

FIG. 117. Principle of Piezo-electric ignition system

crystal (Fig. 117). The element, in this case, consists of a pair of PZT pressure-moulded ceramic rods A and B of $\frac{3}{8}$ in. diameter which are a sliding fit in a hole made in a plastic housing. Between the silvered ends of the two rods is a steel disc, about ·06 in. thick; on each side there is a thin coating of a conducting rubber. Similar discs are used at the outer ends of the rods and thrust pads. One pad is fixed and the outer one is engaged with the compression lever. Between the ends of the housing a high voltage pick-up is mounted. The outer ends of the rods are earthed and the H.T. current generated on compression bridges a small gap between the pick-up and the edge of the central disc. The rubber coating on the disc ensures uniformity of pressure over the rod ends.

There is a simple relationship between the stress applied to the element and the induced voltage, namely, a linear one, so that the voltage is proportional to the applied stress; within the stress region of approximately 7,000 and 8,000 lb. per sq. in., the voltage developed is about 14,000 to 19,000 volts. It is considered probable that the material will withstand higher stresses than 10,000 lb. per sq. in. and also cycles of stress application, before failure, exceeding 100 million.

Tests made by Crankshaw and Arnold* show that the relationship between the ceramic element energy output and voltage is represented by a graph of the form shown in Fig. 118.

Reverting to the practical aspect of this system, pressure is applied to the elements by means of a cam—mounted on the valve-operating camshaft of the engine—which actuates a double-lever mechanism giving a multiplication ratio of 15 : 1. This provides a load of 80 lb. on the relatively small contact area of the element and produces a voltage of about 20,000 for ignition purposes.

The timing of the electrical discharge is not effected by the lever-actuating cam but by an adjustable cam-operated timing

* *S.A.E. Journal*, 1961.

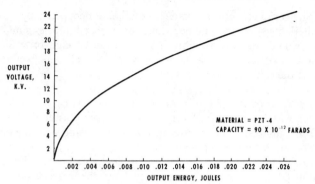

FIG. 118. Relationship between ceramic-element energy and voltage

switch located between the pick-up and the sparking plug cable. The electrical charge which is generated on the release of the compression load is not utilized but is 'earthed' by the timing switch.

The element has been shown capable of producing sparks at 30,000 volts, but the lower values are employed to avoid excessive loading of the element.

The voltage attained on compression of the element is almost independent of the actual rate of load application, so that fully satisfactory sparks are obtained at starting engine cranking speeds as well as at very high speeds. In this connection it is claimed that efficient sparks are obtainable at 10,000 cycles per min., the voltage build up being so rapid that even if the sparking plug is fouled with deposits, an effective spark can still be obtained; as previously stated this is a special advantage, with two-cycle engines.

The system described has been developed by the Clevite Corporation of America and has been applied successfully to industrial engines. The element should operate efficiently for 1,000 hours of continuous service on engines running at 3,000 to 3,600 r.p.m.

The PZT ceramic element has a high energy conversion efficiency, namely, of the order of 70 to 80 per cent under suitable operating conditions.

The complete unit is relatively small, and occupies only a very limited space on the engine; thus, for a 3·25 b.h.p. engine, the unit would measure roughly $3 \times 1\frac{1}{2}$ in. and weigh only 8 ozs. The cost of replacement of the piezo-electric element would be about the same as for a sparking plug.

The Ring Ignition Process. Although non-electrical this method, which was developed for German aircraft, using piston engines, is of some interest, since its use results in dispensing with the usual electrical ignition equipment entirely.

In this method ignition of the air-fuel charge in the cylinder is produced by spraying a special fuel, at the correct moment, which would almost immediately ignite the compressed charge. Such a fuel was mentioned by Dr. Rudolf Diesel in 1898, and subsequently developed by Bayerische Motorwerke of Munich in 1942.

The liquid used is of the ether group, namely, butadiol diethyl ether, or diethyl diglycol ether; it was commonly referred to as 'R-fluid'.

This liquid is not affected to any extent by changes of engine compression ratio and will operate with ratios of 7 : 1 upwards. Further, it will ignite satisfactorily mixture ratios from the leanest to the richest.

The liquid reduces the knocking tendency in fuels to an appreciable extent.

The timing of injection of R-fluid does not appear to be so important as with electric ignition systems, i.e. small changes in injection timing do not affect the rate of pressure rise.

Engine temperatures are considerably lower with this method and no difficulties occur at any rate of engine speed, although the timing of injection must be advanced at higher speeds.

Engine starting difficulties occur, however, so that a set of auxiliary sparking plugs was found necessary; this, of course, necessitates the use of a battery and coil system.

CHAPTER 6

THE MAGNETO

THE magneto, which was universally employed on automobiles and motor-cycles since the introduction of the well-known Bosch model, prior to 1912, established a wide reputation for reliability and efficiency until the 1930's when it was gradually replaced in this country by the battery-and-coil ignition system which had originated earlier in the U.S.A.

Today, the coil ignition system is the standard equipment for production model cars, the magneto being used only for certain tractors, motor cycles, aircraft piston engines, lawn mowers and stationary engines. For high performance engines, such as those of racing cars alternative ignition systems are becoming available, so that automobile interest in the magneto may be regarded as academic only. Under these circumstances it is proposed merely to give a brief outline of the principles and types of magneto that have been used to date.

Principle of the Magneto. The magneto operates on the same principle as the coil ignition system except that it contains its own current generator instead of using a battery for the current supply. Similarly to the coil ignition system it uses a contact breaker, H.T. distributor, condenser and a hand or automatically controlled ignition timing device.

The Bosch and other magnetos of its period employed the revolving armature method of producing the low voltage current. The armature was driven at one-half engine speed for four-cycle engines and at engine speed for two-cycle ones, and it rotated between the poles of a permanent magnet. The coil consists of both a primary and secondary coil winding—corresponding to the windings in the H.T. coil unit as used in conventional coil ignition systems—wound on a soft iron laminated armature.

Fig. 119 illustrates the components of the Bosch-type rotating

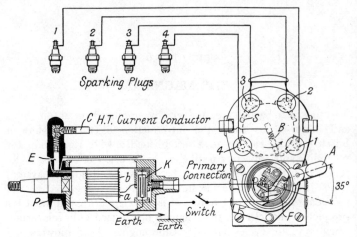

Fig. 119. Typical four-cylinder rotating armature magneto (Bosch)

armature four-cylinder magneto which is driven at engine speed at the tapered-end shaft on the lower left hand side. The primary and secondary coils are indicated at a and b and the condenser across the contacts at K. One end of the primary coil is connected to one end of the secondary coil, the other end of the primary coil being earthed. The other end of the secondary coil is connected to a brass slip ring P, upon which a spring-loaded carbon brush E bears. The H.T. current is collected by this brush and taken to another carbon brush C which makes contact with the distributor rotor brush B that distributes the current to each of the four distributor metal segments, to which the sparking plug

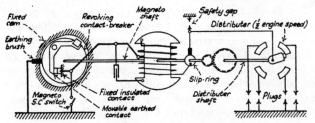

Fig. 120. Schematic layout of four-cylinder rotating armature magneto

THE MAGNETO

cables are attached, as indicated. The contact breaker unit is connected to the common end of the primary and secondary coils and it is driven by the armature shaft. The contact breaker revolves within a cam ring which has two lobes F which produce two 'breaks' per revolution of the armature shaft. The distributor brush B is geared down to run at one-half the armature shaft speed; there are thus two 'sparks' per revolution to suit the four-cylinder engine requirements.

The magneto embodies a *safety spark gap* to provide an escape path for excessive voltage sparks. An ignition switch is provided in the magneto circuit, such that by earthing the primary circuit, the engine will cease to operate.

Fig. 121 illustrates the changes in the flux between the magnet poles and the armature as the latter makes one revolution.

The induced voltage in the primary coil is a maximum for Position (4) and falls to zero in Position (2). It will be observed that there are two maximum and zero positions every revolution.

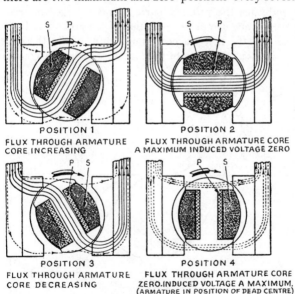

POSITION 1
FLUX THROUGH ARMATURE CORE INCREASING

POSITION 2
FLUX THROUGH ARMATURE CORE A MAXIMUM INDUCED VOLTAGE ZERO

POSITION 3
FLUX THROUGH ARMATURE CORE DECREASING

POSITION 4
FLUX THROUGH ARMATURE CORE ZERO. INDUCED VOLTAGE A MAXIMUM.
(ARMATURE IN POSITION OF DEAD CENTRE)

FIG. 121. Flux changes during revolution of the armature

198　MODERN ELECTRICAL EQUIPMENT

The primary circuit includes a (revolving) contact breaker connected in series and the circuit is completed through a switch and the metal frame of the magneto and the engine—which form the earth return path for the current.

The secondary coil, wound around the primary coil, is in series with the sparking plug selected by the distributor brush or jump spark gap, and has an earth return *via* the outer metal shell of the sparking plug.

When the armature coil unit is in one of the Positions (4) the circuit of the primary coil is broken by the mechanical opening, by means of an internal cam, of the contact breaker points. This causes a sudden collapse of the magnetic flux which induces a high voltage in the secondary circuit, its value being approximately proportional to the ratio of the number of windings in the secondary and primary circuits.

The induced electromotive force and also the current in the magneto is alternating in character as will be seen from Fig. 122, which indicates the primary current and voltage and also the flux wave for one complete revolution of the armature. It will be observed that the flux reaches a maximum, but in the opposite sense, twice per revolution of the armature. When in Position (2)

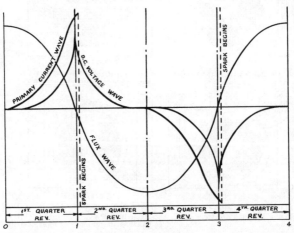

Fig. 122. Flux, voltage and current changes during armature rotation

it is a maximum and no flux change occurs, so that the induced E.M.F. (voltage) is a minimum. Similarly, in Position (0), the E.M.F. is a minimum but of opposite polarity. The maximum voltage occurs at (1) and again at (3), when the flux is changing at its greatest rate, although it is zero at these two positions.

It is therefore arranged for the primary current circuit contacts to separate very near to these two zero flux positions which also correspond to when the voltage change is most rapid.

The fact that the contacts open beyond the zero flux position is due to the distortion of the magnetic field, or flux in the direction of the armature rotation, so that the zero flux is displaced accordingly.

It should be mentioned that the condenser which is connected in shunt with the contact breaker points also plays an important part in the production of the rapid rise of voltage in the secondary circuit, when the contacts open.

In a typical rotating armature magneto the induced H.T. voltage for a magneto speed of 100 r.p.m. may be as high as 10,000, producing a discharge current of ·08 amp. The primary current when the contacts are closed reaches a maximum value of 3 to 5 amps., corresponding to a magneto flux through the armature of 25,000 to 50,000 lines of force.

Timing control of the ignition spark is effected by moving the internal cam ring towards the rotating contact breaker to *advance the timing* and in the direction of rotation of the contact breaker to retard the timing. It will be apparent that the characteristics of the magnetic field and armature will alter for different timing positions.

It is not possible to obtain one optimum timing or 'contact break' position for all positions of the timing lever or engine speed by the hand-control method, but it is usual to adjust the timing so that the optimum break occurs at the full advance position of the timing lever when the magneto is running at low speeds.

The effect of ignition timing for full advance and retard is shown by the graphs in Figs. 85 and 123; the equivalent voltages expressed in three-point spark gap distances for both the magneto and a coil ignition system are shown in this diagram.

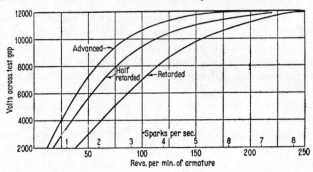

FIG. 123. Effect of magneto ignition timing on the voltage output

It may be mentioned that the later magnetos were fitted with an *automatic centrifugal control* to provide the optimum break conditions at all positions of advance and retard, thus making the magneto equivalent to the coil in this respect.

The later magnetos utilized the more recent alloy magnet steels giving relatively high magnetic characteristics, e.g. coercive force, remanent flux density and (BH) $_{MAX}$ values; such steels include nickel-aluminium steel and 35 per cent cobalt steel. The other alloys used include Alnico, Alcomax, Ticonal, etc. The use of these alloys has resulted in a marked reduction in the dimensions and weights of the permanent magnets and the indefinite retention of their magnetism. Further information on this subject is given later.

Other Types of Magneto. The rotating armature magneto which was widely used over a long period of years was later challenged and to a large extent replaced by other types including (1) *The Revolving Magnet type*, which led to the development of the camshaft magneto and (2) the *Polar Inductor Magneto*.

Revolving Magnet Magneto. In this type the coil member is fixed and the magnet revolves as shown, schematically, in Fig. 124. The magnetic circuit is indicated by the dotted arrowed lines. The coil and condenser are mounted on a detachable part of the magnetic circuit, named the *core*. The contact breaker unit is

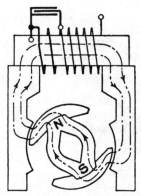

Fig. 124. Principle of the rotating magnet magneto

stationary as in the coil ignition system, the breaker arm being operated by a two-lobe engine-driven cam. It is now possible to use a small magnet unit, due to the excellent properties of modern magnet alloys. The illustration, Fig. 124, shows a two-spark magneto arrangement.

The constructional features of a typical rotating magnet type magneto, for a six-cylinder engine are shown in Fig. 125. It will be observed that this magneto resembles the rotating armature type, but the coil, condenser and contact breaker units are stationary and therefore are not subjected to mechanical stresses due to rotation. The magnet and contact breaker cam, with the distributor rotor arm, the only revolving members, are relatively robust in construction.

Since the contact breaker unit is stationary, it is readily accessible for examination of the contacts and other parts. It is operated by a cam similar to that used in the coil ignition system. The number of lobes on the cam is equal to the number of sparks required per revolution of the magnet drive shaft, i.e. to the number of magnet poles alternately, by using two contact breakers the cam will need only one-half the number of lobes.

The Camshaft Magneto. This high performance rotating magnet magneto (Fig. 126) is driven by a vertical or inclined shaft in an identical manner to the coil ignition distributor drive

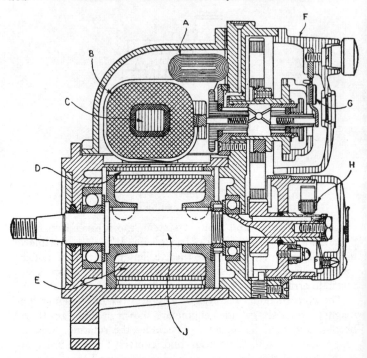

Fig. 125. A typical six-cylinder rotating magnet magneto. (A) condenser. (B) windings. (C) laminated armature core. (D) laminated magnet pole. (E) ring magnet. (F) distributor. (G) distributor brush holder. (H) contact breaker. (J) straight-through shaft

shaft and it can replace the Lucas coil ignition distributor contact breaker unit. The rotor which runs on two ball-bearings is built up with a single circular magnet with four or six laminated pole shoes according to the type of magneto. The magnet and pole system is arranged so that the number of sparks produced per revolution of the magnet shaft is equal to the number of engine cylinders. The shaft is driven at engine camshaft speed, i.e. one-half engine crankshaft speed, so that the usual internal gearing of the magneto is dispensed with. A high quality magnet nickel-aluminium alloy, Nifal or, alternatively, Alnico, is used for the rotating magnet.

THE MAGNETO

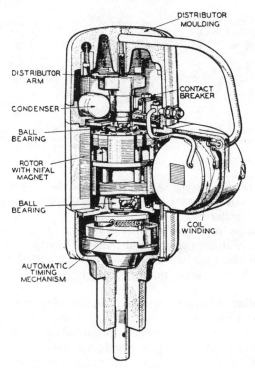

Fig. 126. The camshaft magneto (Lucas)

The contact breaker is of standard design, as used on coil ignition units, but usually with platinum contacts. The drive is taken through an automatic advance and retard mechanism to the rotor at one end of which is the contact breaker cam. The same arrangement of contact breaker and distributor, as used in coil ignition units, is employed.

The ignition timing is advanced automatically, by a centrifugal device (Fig. 127) as the engine speed increases and retarded as the engine slows down. The advance mechanism is housed in the lower part of the body between the rotor and driving shaft. It is arranged at low speeds that an auxiliary centrifugal weight acts with a main weight but at high speeds the auxiliary weight acts

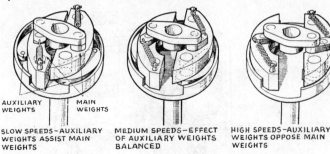

FIG. 127. Illustrating the automatic timing advance mechanism

against the latter weight. By this means the weights can be arranged to give a timing (ignition advance–speed) graph of any desired characteristics.

The Lucas magneto illustrated in Fig. 126 gives a very good performance at low engine speeds, namely, from 30 to 50 r.p.m., so that starting requirements are met, while it will operate satisfactorily at speeds up to 10,000 r.p.m. (engine).

The Polar Inductor Magneto. This type of magneto has both the coil unit and magnet stationary and employs a rotating soft-iron member having a number of projections at equal intervals around its central part. These members are known as inductors and are arranged as shown in Fig. 128 which shows the four-spark magneto system.

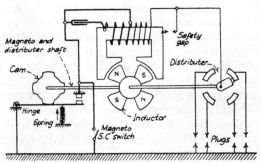

FIG. 128. Principle of the polar inductor magneto

THE MAGNETO

The main flux member is shaped like a hollow metal rectangle with a curved portion, below, cut away. The curved ends of the magnet form the poles and these induce opposite polarity poles in the lobes of the inductor as they rotate close to the poles. The rotation of the inductor between the magnet poles causes changes of magnetic flux, as shown in Fig. 129 which gives the flux changes

POSITION 1.
FLUX THROUGH WINDINGS.
A MAXIMUM.

POSITION 2.
FLUX THROUGH WINDINGS
ZERO.

POSITION 3.
FLUX THROUGH WINDINGS
A MAXIMUM (reversed)

FIG. 129. Showing the flux changes during rotation of the magnet system.
(B.T.H.)

for three positions of the inductor. Four small air gaps are interposed in the complete magnetic circuit, one being arranged between the annular end of each inductor and the ring magnet pole which surrounds it and also a gap between the end of the induction lobe and its corresponding pole. The magnet system consists of two alloy magnets, the ends of which make contact with two laminated iron rings.

During each revolution of the polar inductor there are four complete reversals of flux giving the four sparks per revolution, so that for a four-cylinder four-cycle engine the inductor (and contact breaker cam) shaft will revolve at one-half engine speed.

When the inductor rotates, an alternating current is generated and it reaches a maximum value four times per revolution. The inductor lobes, which move past the poles of the laminated circuit carrying the armature windings become alternately N and S poles and as each pair of lobes is followed by a second pair the flux in the laminated circuit is reversed. The primary circuit is broken at the proper moment—as fixed by engine timing requirements—by means of an orthodox type of contact breaker,

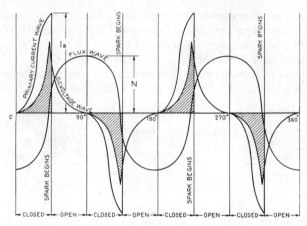

Fig. 130. The flux, primary current and voltage changes during four cycles of the inductor (B.T.H.)

so that a high voltage spark occurs in the secondary circuit across the sparking plug electrodes.

Fig. 130 shows the manner in which the flux, primary current and voltage change during the four cycles, constituting one revolution of the inductor.

This type of magneto is particularly suited to multi-cylinder engines, including eight- and twelve-cylinder vee-type engines, since four sparks can be obtained for each revolution of its drive shaft.

The principal advantages of the polar inductor magneto include low operating speed, the use of a straight through driving shaft, stationary coils, magnet and contact breaker unit, absence of H.T. pick-up brushes, ample leakage prevention surfaces and a rotary type of safety spark gap. The magneto is relatively simple to dismantle for inspection or servicing purposes.

There are other satisfactory designs of magnetos having stationary combined primary and secondary coil units. A typical example is the sleeve-inductor type in which two iron segments, each subtending an arc of about 90° revolve around a stationary H-shaped armature.

Notes on Magnets. With its development over a long period of years the magneto is practically as good as the coil-ignition unit in providing good ignition sparks at low engine speeds while at high speeds the later 'four-spark-per-revolution' magneto and the camshaft magneto provide fully satisfactory ignition at the maximum operating speeds.

The stationary armature, condenser and contact breaker unit types of magneto are extremely reliable and can produce powerful sparks at the higher engine speeds but should the sparking plug electrodes show indications of burning, suitable shunts to prevent this effect can be fitted to the magneto.

The magneto, being a self-contained engine-driven unit is independent of the vehicle battery and electrical wiring system and is therefore more reliable on this account.

The magneto's main disadvantage is that it is, relatively, much more expensive to manufacture than the coil-ignition components.

It may be mentioned that, apart from the low speed spark improvement of the modern magneto, it can be fitted with a simple device, known as an *impulse starter*, which is interposed between the engine-drive coupling and the magneto shaft. By means of a spring and cam and pawl release device, the magneto shaft can be given a sudden impulse which greatly accelerates the shaft and thus the armature or rotating magnet, so as to provide a powerful spark at a low cranking speed.

Magnet Metals. The present alloy steels used for permanent magnets have been developed over a period of years from the original heat-treated, high carbon steels. The first major development was the tungsten steel magnet, with about 5 to 6 per cent tungsten, which was followed by chromium, cobalt-chrome and high cobalt steels. The more recent alloys, which possess excellent magnetic properties belong to the iron-nickel-aluminium group, sometimes with cobalt. Typical alloys of this group include Alni Alnico, Alcomax and Ticonal.

The properties of various magnet steels and alloys are given in the accompanying table, while the marked advance in the development of magnet steels is illustrated by the sizes of some typical magnet steels, shown in Fig. 131. The amount of magnet

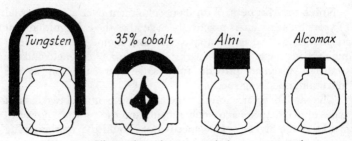

FIG. 131. Illustrating advances made in magnet metals

Properties of Permanent Magnet Metals (Average)

Steel or Alloy	Coercive Force (oersteds)	Remanence (flux density) Gauss	BH (max)*
1% carbon steel	55	9,000	200,000
6% tungsten steel	67	11,000	275,000
3% chromium steel	65	9,000	250,000
15% cobalt steel	183	8,500	600,000
35% cobalt steel	250	9,300	850,000
Nickel–aluminium–iron	505	6,000	1,000,000
Alnico	570	7,350	1,700,000
Alcomax	480	12,700	3,550,000
Ticonal	485	13,600	4,300,000
Alcomax II	510	13,500	4,000,000

* B—magnetic flux density.   H—magnetizing force.

metal for the same magnetic circuits, for a revolving armature magnet are indicated by the black (magnet) areas, in each case.

It will be obvious that with modern magnet metals the size and weight of a magneto can be reduced considerably, in comparison to earlier materials. More recently still further advances have been made in magnet-alloys by the introduction of Columax, Alcomax III and Alcomax IV, Hynico, etc. The remanence values have been extended to 13,500 gauss and the coercive force to between 650 and 900 oersteds. Still further reductions in volume and weight have thus been obtained; for example the Columax magnet is about one-half the dimensions and weight of the Alcomax II one.

THE MAGNETO

The Flywheel Magneto. While operating on the same principle as the orthodox vehicle magneto, the flywheel-type makes use of stationary armature coils and rotating permanent magnets rotating at engine speed. The H.T. and L.T. overwound coil unit, as well as the contact-breaker and condenser are mounted on a fixed circular disc type aluminium plate. The aluminium flywheel is of three-spoked form in which a pair of similarly-shaped curved magnets are screwed to the inside of the flywheel rim, as indicated in Fig. 132 at K. The H.T. coil unit is denoted by C and the contact breaker by D. The sparking

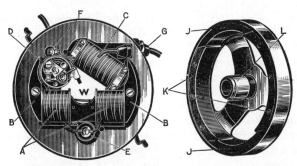

FIG. 132. Combined flywheel magneto and lighting dynamo. (A) lighting coils. (B) fixed plate. (C) H.T. coil unit. (D) fixed contact breaker. (E) lighting cable connection. (F) L.T. lead connector. (G) H.T. cable to sparking plug. (J) flywheel. (K) magnets. (L) flywheel boss and cam. (W) contact breaker arm heel

plug cable is shown at G. The magneto unit illustrated is designed for use on light motor cycles of the two-cycle kind and it is provided with a pair of additional coils A, which are interconnected and used for lighting current supply purposes. The magnets are arranged with their two like (N or S) poles adjoining.

As the magnet system rotates a magnetic field or flux through the coils is created and this flux is constantly changing during each revolution of the magnets. The flux alternates in the positive and negative sense and, as in the ordinary rotating armature magneto, the contacts are arranged to open when the primary current induced in the H.T. coil unit is a maximum.

The contact breaker moving arm is provided with a fibre heel, marked *W* in Fig. 132 (left hand), which bears on a cam *L* formed on the boss of the flywheel unit, which is secured to the engine crankshaft. In order to alter the timing of the ignition the back plate is slotted so that it can be rotated through a small angle; the fibre heel can therefore be varied in its position, relatively to the cam.

In another design of flywheel magneto the two lighting coils are on opposite sides of the cam and the H.T. coil is connected between the ends, so as to form a U-shaped member. The flywheel magneto can also be made with fixed armature coils, H.T. coils and magnets, but with a rotating inductor.

The principal advantages of the flywheel magneto are: (1) Simplicity. (2) Provision of a strong spark at low flywheel speeds, and (3) Constant spark over the full range of ignition advance and retard. The lighting section provides an alternating current which increases in value as the engine speed rises; but the normal intensity is not comparable with that of battery lighting.

Condenser-Discharge Flywheel Magneto. Designed to overcome starting and slow-running difficulties of small two-cycle engines this ignition system produces a high voltage discharge at the sparking plug points, which should enable sparks to be produced on dirty plug electrodes. It operates on the principle of employing a feed coil producing A.C. current which is converted to D.C. current by a small rectifier to charge a condenser. This charge is retained until the *contact breaker points close*, when the condenser discharges through the primary of an H.T. ignition coil, thereby creating a high voltage spark in the secondary circuit. It will be seen that the spark occurs when the contacts close; this is the opposite to that of the usual magneto and coil ignition systems.

It is claimed that dirty contacts do not affect the operation of the system and that there are greater permissible contact point clearance variations than is ordinarily the case. Experiments with this system indicate that it is necessary to retard the ignition by 3 to 4 degrees, due to the higher flame rate which is induced; the maximum output of the engine is not affected by this ignition retard.

CHAPTER 7

THE SPARKING PLUG

Some General Considerations. The successful operation of the sparking plug, as discussed in Chapter 4, requires certain specific electrical conditions, including satisfactory ionization of the electrodes gap, the correct capacitance and inductance spark component energies, rapid rates of secondary circuit voltage rise and, for the higher engine speeds, namely, those above about 7,000 r.p.m. ignition systems that will allow the extremely high sparking rates; these are of the order, 375 to 500 sparks per sec. in modern higher speed six-cylinder engines.

From the theoretical and experimental aspects it will be evident that the sparking plug has to withstand, for long periods, severe operating conditions. These include the following:

(1) High peak pressures of an impulsive nature in the combustion chamber, of the order 650 lb. to 800 lb. per sq. in., but higher peak pressures may result.
(2) High operating temperatures in the vicinity of the plug in its combustion chamber; these may attain maximum values, during combustion of 2,000° to 3,000° C.
(3) Gas leakage possibility due to the joints between the insulator and plug body.
(4) Erosion of the electrode points due to the passage of the high voltage sparks.
(5) Corrosion effects, more especially when leaded (premium) fuels are used over appreciable periods.
(6) Exposure to fouling of the insulator end portion by oil and mixture combustion deposits.

The sparking plug metal shell must be strong enough to resist distortion due to screwing it into the combustion chamber and also unscrewing it. The insulator must withstand normal handling and usage without fracture above the metal shell—a fault liable to occur with the earlier porcelain insulated plugs. In this connection it may be noted that the recommended *tightening torques* for standard 14 mm. plugs is from 25 to 30 lb. ft.

Sparking Plug Materials. The components of a typical modern plug, shown in Fig. 133, include the electrodes, one being

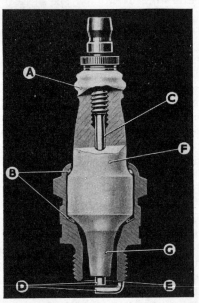

Fig. 133. Typical modern sparking plug (A.C.). (A) insulator (top). (B) sealing washers. (C) central H.T. electrode. (D) exposed (sparking) electrodes. (E) earthed electrode. (F) insulator at metal shell. (G) insulator, portion exposed to cylinder gases (unglazed)

fixed to the plug shell and the other fixed to the central conductor to the other end of which the H.T. cable from the distributor is attached; the electrodes across which the ignition sparks occur; the insulator unit and the seals which render the insulator gas-tight in the plug shell. In addition there are the brass terminals, which are screwed to the metal conductor member.

The plug shell is usually machined from a low carbon free machining steel, such that apart from mechanical strength considerations it must withstand rolling over at its upper end to make sound joints at the top and bottom joint faces of the insulator (Fig. 133).

Plug electrode materials have been the subject of experiment in the past; also their proportions and shape. In regard to electrode material this, as stated earlier, must give the maximum plug life under spark erosion action and corrosive action due to the combustion products. The high temperatures to which these electrodes are exposed, necessitate the use of high melting point metals or alloys, e.g. certain nickel-containing alloys which are also good electrical conductors.

In this connection nickel-chromium alloys are noted for their corrosion resistance, but the addition of small proportions of other elements, such as *barium*, improves their 'work function', i.e. renders more uniform the sparking voltage; this is known as *isovolt alloy*. Nickel, alone, has proved satisfactory for plug electrodes.

It has been shown that *platinum* or one of its alloys with tungsten or iridium, is the best metal for electrodes, since it has an excellent resistance to erosion and corrosion and reduces, appreciably, the working voltage. Sparking plugs with platinum electrodes have a considerably longer life, namely, at least three times that of nickel alloy electrodes. It is also known that whereas a nickel electrode is subject to pre-ignition at 700° C, a platinum electrode can be raised to 1,200° C before pre-ignition occurs. Usually, the platinum central electrode is of wire, about ·02 in. diameter and ·10 long. The recommended gap is ·020 to ·025 in.

It may be mentioned that satisfactory alloys with nickel, for plug electrodes can be made with one or more of the following elements, namely, manganese, chromium, tungsten and silicon.

Tungsten is unsuitable for plug electrodes, despite its excellent erosion and corrosion resistance, due to its high sparking voltage and also brittleness, which prevents electrode bending for gap adjustment.

Sparking plug insulators were originally of porcelain made with different kinds of silicates but these were liable to troubles, such as fractures due to their brittleness, low voltage breakdown properties, and relatively poor thermal shock properties, i.e. resistance to intermittent heating and cooling effects.

Mica Insulators. To overcome these troubles mica-insulated plugs were introduced. In their construction, mica strip was wrapped around the central conductor and mica washers threaded over, afterwards being compressed axially and machined to the desired external shape and size. These plugs were widely used on ordinary and racing cars, aeroplanes and motor-cycle engines with satisfactory results until the higher octane fuels, containing tetraethyl lead, were introduced. It was shown that the lead products attacked the mica and reduced the working life of the plug, appreciably.

Other disadvantages of mica were those of low thermal conductivity; low thermal expansion, namely, about one-half that of steel, thus making gas-tight joints more difficult; deterioration, if overheated, due to a change in its chemical constitution, whereby, the 'water' element is eliminated, and, finally, the tendency of the mica to increase the capacity of the H.T. secondary circuit of the ignition system.

Improved Ceramic Insulators. Important advances were made in porcelain insulators for aircraft engines, by replacing the quartz and feldspar constituents—which caused both volume changes and insulation efficiency—by inert components, such as *sillimanite* and by finer grinding and higher production temperatures, so that these insulators were much stronger, mechanically, and possessed better electrical and thermal properties. Further, it was found possible to reduce the size of the insulator, so that lighter plugs resulted.

Aluminium Oxide Insulators. The more recent insulating material is that made from aluminium oxide sintered at high furnace temperatures into a hard and uniform material. Since it is difficult, on account of the high melting point of alumina, to employ, commercially, the pure substance it is necessary to mix from 5 to 10 per cent of other substances to alumina. These sintered alumina compositions are known by various trade names but, in general, possess more than three times the tensile and compressive strength of the previously used porcelains; have from five to ten times their heat conductivity; about four to five times their dielectric strength and a considerably increased resistance to thermal shock. Also, such insulators, while not being free from fouling troubles are not attacked, chemically, by leaded fuel products. Another advantage is that cleaning of the insulator surface, when fouled by carbon or oil deposits, is a simple matter, since the use of a sand blast for this purpose does not damage the surface. With the general adoption of sintered alumina insulators, all of the previous troubles, other than that of fouling, have vanished and breakdown of insulators is a rarity.

The insulator, as mentioned earlier, is exposed to very high temperatures at the combustion chamber side and to the relatively low outside air temperatures at the plug terminal end. Therefore, there exists a temperature gradient, under engine operating conditions, somewhat as indicated in Fig. 134 (A). The conduction of

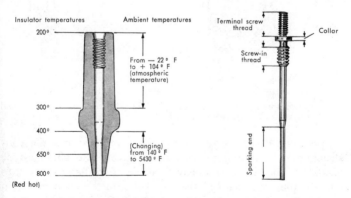

FIG. 134. Sparking plug insulator (A) and central electrode (B)

heat away from the exposed end of the conductor must be provided for so that extreme temperatures above and below the working limits are avoided as is explained later.

Finish of Ceramic Materials. It is known that the lead salts derived from the combustion of high octane fuels have a deleterious action on certain ceramic materials, forming a low melting point silicate glass, having chemical activity and becoming a good electrical conductor at high temperatures. For this reason the ends of insulators exposed to combustion products are left *unglazed* in order to retard the effects of the lead salts. The outer exposed parts of all insulators, namely, at the plug terminal end, are always *glazed* as a protection against moisture, oily deposits or dirt.

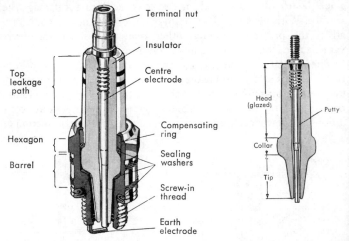

FIG. 135. Showing construction of sparking plug (Bosch)

FIG. 136. Insulator and central electrode assembly

Sparking Plug Construction. Fig. 135 illustrates the construction of a typical modern plug, having the alumina-type insulator. The central conductor is made of steel and screwed to take the plug terminal and also to screw into the insulator down to its collar. The electrode at the lower end (Fig. 134 (B)) is welded to the steel conductor. The central conductor-electrode member is made a gastight fit in the insulator hole with a special composition marked 'putty' in Fig. 136 and with talc powder cemented in, in other instances. More recently the joint is made with a kind of

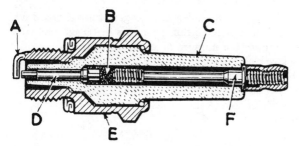

FIG. 137. Sparking plug with alumina insulator, using conducting glass-sealing method (Lucas). (A) earthed electrode. (B) conducting glass seal. (C) alumina insulator. (D) central electrode. (E) metal body, or shell. (F) push-on type cable connector terminal

conducting glass material (Fig. 137) which readily bonds itself to the metal and insulator. In connection with the central metal member this should have the same coefficient of heat expansion as the insulator, to avoid differential expansion adverse effects.

The sealing of the insulator to the metal body of the plug is important since there must be no risk of leakage of gases from the combustion chamber which, otherwise, would not only affect the plug temperature, but to some extent alter the compression and maximum pressure values. The method shown in Fig. 133 is that of using special sealing washers (*B*) on the inclined faces of the insulator. After insertion of these washers the thinner upper end of the metal shell is rolled over under a certain amount of pressure, usually with the shell heated, so that on cooling the contraction effect renders the joint completely gastight.

Sometimes a compensating resilient steel washer or a compensating ring is used to ensure gastight joints at all temperatures —as shown in Fig. 135.

The earthed electrode is welded to the annular end of the metal. In this connection, the previous multi-electrode plugs and other arrangements of single electrodes have been replaced by the hooked earth electrode and central electrode method of producing the sparking gap; this arrangement is shown in Fig. 133.

With multi-electrode plugs while there are alternative paths for the spark, thus giving longer electrode life, each path is an added risk to failure by bridging of the path with oily deposits.

The hooked electrode leaves the spark fully exposed all round the central electrode and thus promotes better ignition conditions for the compressed mixture of air and petrol.

The sparking plug itself is made a gastight fit in the combustion chamber, by means of a copper-asbestos or similar washer which compresses as the plug is screwed down; it is an advantage to use the non-detachable washer which is available for some plugs, e.g. the Bosch.

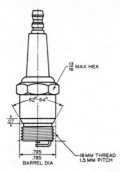

FIG. 138. American Standard sparking plug, with conical sealing seal (S.A.E.)

Certain American engines, e.g. the Ford range, use sparking plugs having conical seating faces which mate with conical seatings machined in the plug screwed holes of the cylinder head, no sealing washers being used. The seating face angle on the cylinder unit is 59° to 60°, while that on the plug (Fig. 138) is 62° to 64°. The seating faces must be quite smooth and concentric to within fine limits, specified by the Society of Automotive Engineers.

Sparking Plug Temperature Ranges. For each design of engine there is a definite range of temperature for the exposed surfaces of the sparking plug at which the plug will operate satisfactorily and remain free from carbonaceous deposits. The combustion chamber design, water cooling passages, compression ratio, plug location and conduction of heat from the plug are principal factors influencing the correct type of plug for each engine.

In this connection there are three important considerations concerning the suitability of a plug in any engine, namely: (1) Pre-ignition. (2) Plug cleaning and (3) Carbon and oil deposition.

(1) *Pre-ignition*. Premature ignition, i.e. ignition of the charge before the electrical ignition spark occurs is known to be due to some overheated part in the combustion chamber, known as a *local hot spot* which becomes heated to redness or incan-

descence by the combustion gases; usually, this spot consists of a projecting piece of metal, or burr, which readily heats up. The electrode of a sparking plug can cause pre-ignition if, due to the combustion chamber conditions, the plug cannot be cooled sufficiently. Further, if the electrodes protrude beyond the surface of the combustion chamber pre-ignition can occur.

If the surface temperature of a sparking plug exceeds 850°, then pre-ignition will usually occur, so that a type of plug having better cooling properties must be used. Commercially, the correct plug for *hot engines*, is known as a *cold plug*; in general this type of plug has a much shorter heat flow path from the exposed insulator surface to the metal of the combustion chamber, as depicted at (B) in Fig. 139.

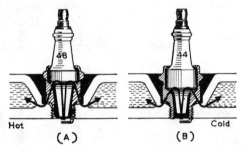

FIG. 139. Types of heat-range sparking plugs. (A) hot plug. (B) cold plug (A.C.)

(2) *Plug Cleaning Property*. The working temperature range of a sparking plug must be sufficiently high to ensure that any carbon deposits on the exposed insulator surfaces are burnt off after they form. This temperature must not, however, be sufficient to promote pre-ignition. The self-cleaning temperature range is known to be between about 500° to 600° C.

(3) *Carbon and Oil Deposition*. These deposits will form on the exposed surfaces of the plug at the combustion chamber end, if the plug is *too cool*, i.e. if its temperature is lower than 350° C. In this case a *hot plug* would be employed, to raise the surface temperature.

Fig. 139 (A) illustrates a typical hot plug of A.C. design which, it will be seen, has a much longer heat path, giving delayed cooling, than that shown at (B).

The sparking plug manufacturers supply plugs with different heat ranges, in each size of plug, i.e. hexagon and screw thread, so that the plug requirements of any engine can be met.

In general the *hot plugs* have a much *longer insulator nose* than the *cold plugs* but the plug temperature is influenced, also, by the insulator size and cross-sectional area, the thermal conductivity of the insulator and the design and metal of the central electrode.

As a result of practical investigations it has been possible to develop and produce, commercially, a range of sparking plugs that will cover the special temperature range requirements of all automobile and motor-cycle engines, satisfactorily. Fig. 140

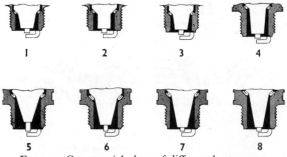

FIG. 140. Commercial plugs of different heat ranges

illustrates the various A.C. sparking plugs available, arranged in the order of their *heat ranges*. In this arrangement the No. 8 plug is the *hottest* one, i.e. is suitable for the coldest running engine, while No. 1 plug is the *coldest* and suitable therefore for the hottest running engine. These illustrations show how the lengths of the insulator tip decrease, with increase of the mean operating temperature of the engine

Sparking Plug Fouling. While the effects of fouling, i.e. deposition of solid matter, derived from the combustion of the fuel and lubricating oil, have been largely overcome by the design of suitable heat range plugs, other methods of dealing with plugs which have definite fouling tendencies have been under investigation.

One of the most promising of these is the previously-mentioned *high frequency* ignition system in which the sparking voltage builds up with extreme rapidity, e.g. less than one-millionth second, and is able to produce an ignition spark at the plug when it is badly fouled. Thus, when the electrical resistivity is of the order of 2,000 micromos, the plug will 'fire' with this system, where the ordinary coil ignition system may fail with only 5 micromos. This method has been used successfully for aircraft piston engines but its relatively high cost is the present drawback for automobile engines.

Another method is under investigation, namely, the *series-gap* or *intensified* plug in which there is a built-in gap in the central conductor-electrode unit. It is necessary to employ high frequency ignition with this type of plug. The method of producing precision dimension gaps involves much difficulty, so that here, again, the cost of the system is high.

Effect of Leaded Fuel. The higher octane, or premium grade fuels used in modern engines contain certain additives, the most important of which is tetraethyl-lead. This lead is converted to lead oxide which is a solid at combustion temperatures.

In order to prevent the formation of this deposit, ethyl-dibromide or ethyl-dichloride is added to the fuel so that the lead content is converted into lead bromide or lead dichloride, which are both gaseous and are discharged from the cylinder with the exhaust gases. In practice, however, this chemical process is not fully completed, with the result that lead oxide and sulphate, which are solids, are left behind to form deposits on the plug electrodes and exposed insulator tip, on the valve heads and combustion chamber walls (Fig. 141). The deposits have also a corroding tendency and can attack the electrodes and certain insulator materials, as indicated in Fig. 142 which shows two electrodes badly corroded by leaded fuel. While the deposits are of a powdery nature at lower temperatures they melt at the higher cylinder temperatures and form a hard coating and, above 400° C, become electrical conductors.

Fig. 141. Lead salt deposits on sparking plug

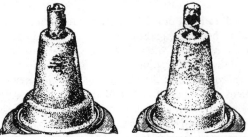

Fig. 142. Effect of lead salt deposits attacking electrode and insulator tip of sparking plug

By using the correct heat range sparking plugs, this deposition can be minimized, but in the case of certain cool plugs deposition is accelerated.

When leaded fuels are used the deposits on the exposed metal of the plug, and its electrode, are of a light greyish yellow colour with reticulated broken line surface markings (Fig. 141). If the insulator deposits are of a powdery nature they can be cleaned by brushing off the powder. Otherwise, sand blast cleaning of the insulator and also the electrodes and other coated metal surfaces is employed.

When iron carbonyl has been used as an anti-knock additive to the fuel this will result in a reddish deposit of a powdery nature on the plug which is easier to remove; moreover, this deposit does not cause corrosion.

The Sparking Plug Gap. The gap or distance between the plug electrodes is governed by the secondary voltage produced by the ignition system and this gap should always be maintained. Thus, in current Lucas coil ignition systems the recommended gap is ·025 in. for engines with compression ratios of 8·0 : 1 to 8·5 : 1. For magneto ignition a somewhat smaller gap, namely, of about ·020, is often employed to ensure easy cold starting.

Some engines, more especially those of American design, operate more satisfactorily with larger plug gaps, e.g. ·030 to ·037 in., a common value for many engines being ·035 in. With a suitably designed ignition system such engines—with their larger capacity cylinders appear to operate much better with these wider gap plugs. The exact reason for this is not altogether established but from indicator diagrams taken from engines using alternatively, wider gap plugs and smaller gap plugs, it was established that the performance with smaller gap plugs showed erratic pressure variations and misfiring occurred occasionally, but with wide gap plugs regular performance and maximum output was obtained. If, however, higher electrical energy was applied to the small gap plugs, more uniform running, but rather less output would be obtained.

Whatever the initial plug gap, it is not possible by electrical or metallurgical means to prevent spark erosion effects which eventually result in the widening of the gap. But if not corrected by re-setting the metal shell electrode—by judicious bending—the voltage requirements will increase and finally misfiring may occur, although only after a long period of service; poor starting will also result.

It is also known that wider gap plugs, e.g. ·030–·037 in. will ignite weaker air-fuel mixtures than smaller gap plugs and that with wider gaps there is a somewhat reduced plug fouling tendency. Smaller gap plugs are the more liable to foul across the electrodes.

When checking the sparking plug gap it is advisable to use a round wire type of gauge rather than a flat feeler gauge, on account of the irregularities in the eroded surfaces which result in a smaller gap reading with the flat gauge.

British Standard Plugs. Sparking plug dimensions have been standardized in the 14 mm. size only.* In this connection the American Ford Company uses the 18 mm. size. Fig. 143 illustrates the automobile, as distinct from the aircraft standard plug, and shows the accepted nomenclature.

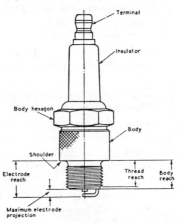

FIG. 143. British Standard 14 mm. sparking plug

The body and electrode reaches are standardized, but the thread reach is variable to manufacturer's requirements.

The standard screw thread is that of the Système International, i.e. the 60° truncated thread form of 14 mm. (0·5512 in.) outer diameter and 1·25 mm. pitch (0·0492 in.).

Three alternative types of terminal are specified of which that shown in Fig. 143 is the most popular. The thread used is the No. 8–32 UNC (Class 2) one.

* British Standard 45 : 1952 *Sparking Plugs*.

THE SPARKING PLUG 225

Plug Polarity. During the erosion or burning process, metal is transferred from the positive to the negative electrode so that as the outer or metal shell or earth electrode is usually larger than the central electrode it is better to make this the *positive* electrode —as is done in the positive earthing system. Moreover, as mentioned before, as the central electrode is the hotter and the plug voltage decreases as this electrode's temperature increases, this is a further advantage of the positive earthing system.

Plug Life. With ordinary maintenance attention every 2,000 to 3,000 miles the alumina-type plug with nickel alloy electrodes has a reliable working life of 10,000 to 12,000 miles, after which it is liable to troubles, e.g. partial short circuiting along the exposed insulator tip, gas leakage, surface cracks on the insulator, etc.

With recent ceramics and platinum-iridium electrodes plug life has been extended to 30,000 to 50,000 miles. In all cases, periodic attention is necessary to ensure that the insulator surfaces and the exposed metal, including the electrodes, are kept clean; the electrode gap should be maintained at its correct value.

Special Types of Plugs. Special plugs are available for enabling the temperature of the insulator and central electrode, under various operating conditions to be measured. Typical plugs which include thermocouples arranged in the central conductor-electrode member, include the Lodge, Champion and Bosch makes, of which the Champion plug is illustrated in Fig. 144.

A combined sparking plug and cylinder pressure sensing element has been used in connection with a high-speed engine indicator, namely, the Dobbie McInnes 'Farnboro' type.

Reference was made earlier in this chapter to the *surface discharge* plug

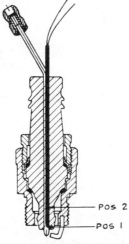

FIG. 144. Combined sparking plug and thermocouple

for an alternative ignition system to the conventional one, and also for use in initiating the ignition of the fuel-air mixture in gas turbines.

When the ordinary high voltage two-electrode plug is used to ignite the fuel in the gas turbine combustion chamber, a series of sparks is discharged across the wider gap of this plug, namely ·05 to ·07 in. by means of a booster coil operating on the trembler coil principle and providing about 60 sparks per minute at a high voltage. The fuel spray in the vicinity of the plug is ignited and the flame spreads through connecting ducts to the other combustion chambers.

An improvement on this method is to incorporate a small fuel atomizer with the electric igniter plug, so that the fuel spray on ignition by the electric sparks is projected as a torch flame, and quickly ignites the main fuel spray; this unit is known as a torch igniter (Fig. 145).*

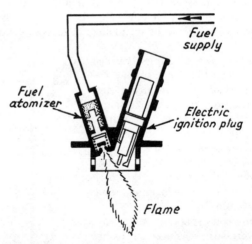

FIG. 145. Combined gas turbine fuel atomizer and electric ignition plug

Interference Suppressor Plugs or Attachments. Special resistor attachments can be fitted to sparking plug terminals to

* *Small Gas Turbines.* A. W. Judge. (Chapman & Hall Ltd., London.)

suppress interference effects upon wireless reception and to some extent, television. The attachments consist of straight or angular moulded caps, having a resistor of 5,000 to 15,000 ohms which connects the sparking plug cable to the plug terminal, thus placing the resistor in series with the H.T. cable conductor. Alternatively, a later improvement embodies the resistor within

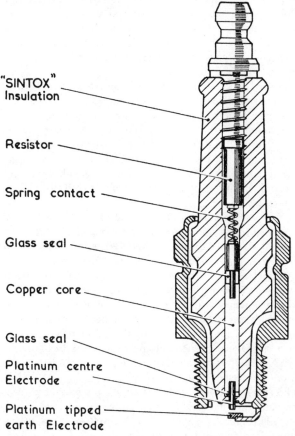

FIG. 146. Lodge sparking plug with carbon resistor, for interference suppression

the central conductor unit of the plug itself, thus avoiding any external plug suppressor attachment.

When the suppressor device is built into the plug it consists of a short cylindrical rod of a carbon resistance material, which is arranged in series with the conductor terminal screw and a lower copper core to which the central electrode is attached. Fig. 146 shows a sectional view of the Lodge plug with the resistor in place. It will be seen that a short spring contact is provided to ensure good electrical connection between the central conductor individual units. A glass seal is provided between the spring contact and the copper core conductor.

Ignition Interference with Television and Radio. It is well known that electrical spark discharges can give rise to serious interference with the performance of radio and television receivers in the vicinity of these sparks. With the very large numbers of motor vehicles and receivers—both on the vehicles and in domestic use—the interference problem became so serious that it was necessary to introduce Government regulations compelling the fitting of interference suppressors to all sources of such interference.

In the case of automobiles, of the various possible causes of interference, that due to *the ignition system* is by far the most important, but the *dynamo, windscreen wiper, vibrator unit of current or voltage control* regulators and the *heater fan motor* can cause interference. In the case of dynamos and motors, the brushes are the sources of spark production. In general, when the current in a low frequency or direct current circuit is varied or interrupted suddenly, it can cause electromagnetic fields which interfere with radio or television operation. In the radio receivers this interference is evident as intermittent clicks or noises superimposed on the normal reception, while in television receivers bright intermittent flashes occur on the pictures.

The spark produced at the sparking plugs is of an oscillatory nature and the energy of the discharge is considerably more than is actually required for ignition of the mixture, so that the surplus energy is radiated in the form of waves which are picked up by receivers within the vicinity of the spark source.

THE SPARKING PLUG

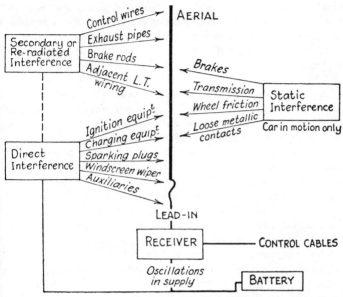

Fig. 147. Principal sources of radio interference, on automobiles

It has been shown that it is the energy stored in the self-capacitance of the sparking plug, the H.T. cables and the secondary winding of the H.T. coil which in discharging across the spark gap at an extremely rapid rate, produces the oscillatory current, some of the energy of which is radiated into space.

The amount of interference depends upon the layout of the ignition wiring system, being less for a compact system with closely arranged components and short cables than with a more widely distributed system.

In regard to the official requirements *the regulations* make compulsory the suppression of interference in Band 1, namely 40–70 Mc. per sec., such that the interference producing field must not exceed 50 microvolts per metre at a distance of 10 metres from the source.

It is, of course, necessary with car radio receivers to fit efficient interference suppressors. In connection with the ignition source of interference it is *important to maintain the ignition system* in an efficient condition since any sources of misfiring, or spark production, e.g. dirty contact breaker points, too wide a rotor gap, faulty conductor connections to cables, etc., can be sources of interference.

Suppression of Ignition Interference. The chief possible sources of interference in a well maintained ignition system are, as follows: The sparking plugs, the rotor arm spark gap and to a lesser extent the low tension contact breaker points.

When the spark is of an oscillatory nature the range of the oscillations can be reduced and their effect reduced practically to a continuous discharge by inserting a resistor of about 5,000 to 15,000 ohms in series with the voltage conducting cable, as close to the spark as possible. Thus, the resistor can be in the form of a sparking plug terminal fitting or it can be embodied in the central conductor unit of the sparking plug itself. In this case each sparking plug has its own resistor.

A more widely-used method is to insert a resistor of 10,000 to 15,000 ohms in the cable between the H.T. coil and the central terminal of the distributor cap. This unit takes the form of a carbon or Morganite resistor inserted in a moulded fitting that screws into the centre socket of the distributor cap, thus placing the resistor between the H.T. cable from the coil and the rotor arm carbon brush contact.

Resistance core cable is often employed—more especially on American cars—instead of individual resistor units, so that the high resistance is distributed along the cable length, e.g. in the H.T. cables to the sparking plugs. In other instances the resistor forms a combined unit at the end of the cable, but is built into the cable. In another design of suppressor cable there is an end capacitance section, provided with a screen, so that electrical impulses are diverted to earth.

Inductances may be used to suppress interference, by placing these or choking coils on both sides of the disturbing source, and by shunting the source with a condenser. *The effect of inductance* is to allow low frequency or direct currents to flow freely but to oppose high frequency currents. *The effect of capacitance* is to

THE SPARKING PLUG

allow high frequency currents to flow but to oppose the flow of low frequency currents or direct currents. A suitable filter-type suppressor of the inductance-capacity kind will remove undesired electrical impulses superimposed on direct current that is flowing in the circuit and allow most of the impulses picked up to be by-passed by the condenser.

So far the suppressors of the resistor type have been considered for the sparking plugs and the distributor cap H.T. cable connection. While in most instances these resistors will suppress any interference effects, it may be necessary in severe cases, to fit resistors not only at the sparking plugs but also at their H.T. cable connections to the distributor cap terminals or sockets.

Another method of preventing interference of the ignition system with the car radio and also domestic radio and television receivers, is by *screening* the various members of the ignition system, by means of braided metal-covered cables with earthing of the braidings. This method, however, has the disadvantage of increasing the capacitance of the complete system, while increasing the cost of the interference prevention installation. The effect of this increased capacitance is to oppose the coil ignition high-speed voltage production.

Interference from the low tension wiring due to its coupling with the high tension system, as in the H.T. coil construction, can be suppressed by means of a condenser of about 0·2 to 1·0 microfarad inserted between 'live' low tension terminal of the H.T. and earth. This small condenser can be mounted on the side of the coil by means of a clip, so that the cable connecting the condenser to the battery (live) terminal on the H.T. coil is kept as short as possible.

In regard to *car radio receivers*, it is usually sufficient to fit a single resistor at the H.T. cable connection, from the H.T. coil, of the distributor cap.

Magnetos may be treated in a somewhat similar manner to coil ignition units, by fitting resistors in the H.T. distributor leads and in more difficult cases separate sparking plug resistors. Vertical magnetos which have external leads between the H.T. windings and the distributor *should not use* the standard resistor that is made for coil ignition H.T. coils, since the energy from the magneto at the higher speeds is considerably greater than from an ignition coil, so that the ignition coil resistor, if used in the main H.T. cable of the magneto would be greatly overloaded.

A more recent type of interference suppressor, introduced by Lucas Ltd. for car ignition equipment is a *new distributor* with a certain measure of interference suppression built in. Here, the carbon brush that conveys the H.T. current from the coil H.T. cable to the centre of the rotor arm is designed to act as a suppressor resistor as well as an H.T. current conductor.

Effect of engine performance. The use of resistors in coil ignition systems does not produce any loss of power from the car engine, if the resistors are of the correct value and quality. It is claimed that such resistors tend to increase the life of the sparking plugs, since the initial oscillating current of each spark is damped out; this, and the reduction of the capacity component of the spark reduces erosion effects at the electrodes of the sparking plugs.

Other Sources of Radio Interference. Mention has been made of interference emanating from the dynamo or any motors on the vehicle; also the vibrating contacts of voltage, or current and voltage regulators, since spark discharges occur at these contacts. It is not difficult to fit capacitors between the insulated or 'live' terminals and earth in certain instances, but these should not be used as shunts across the contacts.

Certain *makes of motor tyres* develop *static electricity*, and when this is discharged sparks occur which give rise to interference. This trouble has been overcome by the use of conducting filler in the composition of the tyre, e.g. graphite; also, in most American car tyres by static collectors which are usually fitted in the front wheels.

CHAPTER 8

THE AUTOMOBILE BATTERY

THE purpose of the battery is to store electrical energy to provide a supply of current for operating the starting motor and other electrical units, e.g. the side and tail lamps, panel lamp, heater unit, radio, etc., when the engine is not working, that is to say, when the dynamo is not charging the battery.

The battery also serves to provide any additional electrical supply, when the car is being driven but the dynamo is not supplying sufficient current to the battery, e.g. when the car is used at night, with its complete lighting system, heater, radio or other additional electrical items in use. This condition occurs in the case of some cars which make excessive demands on the electrical supply which the dynamo cannot cope with, fully; in such cases the red indicator light on the panel 'glows' continuously.

There are two principal types of battery, namely, the *Lead-Acid* battery that is the standard type for automobiles, and the *Alkaline cell* battery that is used largely on heavy vehicles and some motor cycles.

The Lead-Acid Battery. The principle of this battery is illustrated schematically in Fig. 148 which shows a pair of flat plates of which one, designated the *positive* plate consists of a grid made of hard lead-antimony alloy, having a large number of rectangular openings which are filled with *lead oxide* active material, which gives the plate its characteristic *dark brown* appearance. This material consists of very fine particles which give the plate appreciable porosity.

The negative plate is of similar basic construction but its openings are filled with a grade of lead known as *spongy lead* which is very porous; in addition the lead embodies a substance to prevent it contracting and returning to the solid state.

Between the positive and negative plates an inert separator

plate is fixed, so that the plates cannot touch one another and give up their stored energy. The separators are made of a non-conducting material and are of a porous nature. Various materials, e.g. porous rubber, chemically-treated wood, perforated cellulose, rubber, etc., are used for the separators. The plates are immersed in sulphuric acid solution, termed the electrolyte.

In practice several positive and negative plates are used, with separators between each pair of unlike polarity plates. The positive plates are connected together above the level of the electrolyte to form a single unit; the negative plates are also connected together.

Assuming the cell of a battery (Fig. 148) is in the fully charged

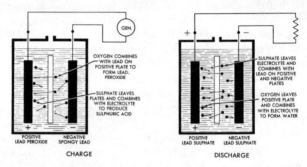

FIG. 148. Illustrating chemical changes during battery charging and discharging operations (Ford-Mercury)

condition, if the positive and negative plates are connected in series with a suitable resistance, there will be a gradual electrical discharge of the cell's stored energy through the resistance in the form of heat energy and during this process the lead peroxide of the positive plate and the spongy lead of the negative plate are converted to lead sulphate, the source of this sulphate being the sulphuric acid solution (Fig. 148). Eventually, with the completion of the sulphate formation no current will flow, so that the cell is then said to be *discharged*, or exhausted.

In order to restore the cell to its original charged state, a source of electrical energy, namely, direct current from a dynamo (Fig. 4) is connected to the plates, as indicated, so that like poles of the dynamo and battery are connected together. When the charging current is switched on, a chemical process commences, which results in the sulphate of the positive plate becoming lead peroxide, and that of the negative plate, spongy lead, again.

The chemical actions that occur during charging and discharging are as follows:

(Charging)				(Discharging)		
Positive plate	Electrolyte	Negative plate		Positive plate	Electrolyte	Negative plate
PbO_2 +	$2H_2SO_4$ +	Pb	$\rightleftharpoons$	$PbSO_4$ +	$2H_2O$ +	$PbSO_4$

It should be pointed out that only a part of the active material can usefully be converted in this way and that the amount of material chemically changed depends upon how quickly the cell is discharged. Thus, for higher rates of discharge the smaller the amount of converted chemical (into electrical) energy.

Specific Gravity and Voltage of Cell. During the discharge of the cell, the *specific gravity* (density) of the electrolyte falls from its initial and maximum value, of about 1·280, progressively to its minimum value, on complete discharge of about 1·15.

The *initial voltage of a single cell*, after it has been charged and a certain period of time allowed for the voltage to become stabilized, is from about 2·0 to 2·1, although immediately after charging the voltage may reach 2·25. During the charging process the voltage attains values as high as 2·6 to 2·7.

The state of charge or condition of the cell is therefore indicated by the *density* of the electrolyte and the *voltage* across its terminals. In practice the former indication is the more reliable and the widely used.

Battery Construction. The construction of a unit cell, consists of a pair of positive and negative plates with a separator of the type described earlier between them. Separators with ribs are placed with the ribs facing the positive plate, so that there is a greater volume of electrolyte near this plate, to improve the acid circulation.

Since the capacity of a battery depends upon the size and

number of battery plates it is found necessary for automobile batteries to connect a number of positive plates in parallel by welding together with a strip of lead on the top, having an integral battery post or terminal, thus giving within the size of a single plate the equivalent of a plate considerably larger. Similarly all the negative plates are connected together at their tops, but oppositely to the positive plates. The separators are placed between adjacent pairs of plates, after the two sets have been located correctly.

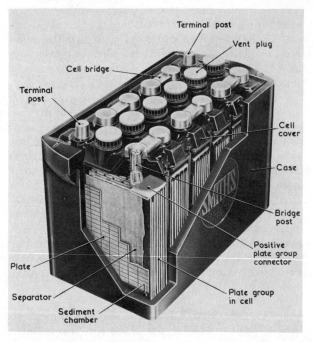

Fig. 149. Typical 12-volt lead acid battery, showing internal construction

Fig. 149 illustrates the construction of a Smith's 12-volt battery and shows also the other parts that constitute the complete assembly. It will be seen that there are six cells, or separate compartments, each with its sets of positive and negative plates, with their separators which, in this case are of microporous rubber; this ensures the free passage of the acid solution and they have a long life.

The battery case is moulded from a tough vulcanized material, e.g. ebonite or a bituminous composition; the separate cell partitions as well as ledges to hold the plates above the bottom, thus forming deposit reservoirs, are moulded in position.

The usual width of plate for automobile batteries is $5\frac{5}{8}$ ins. and height, 4–5 ins.; the plates are from $\frac{5}{64}$ to $\frac{3}{32}$ in. thick; for commercial batteries the thickness is $\frac{1}{4}$ in. The number of plates per cell varies from 6 to 13 in typical car batteries and there is always *one more negative plate* than in the positive set, so that there is a uniform working of all the positive plates.

The number of plates per cell determines the capacity of the cell while the number of cells, in series, governs the battery terminal voltage. Thus, a 40-ampere hour 12-volt battery would have 6 or 7 plates per cell and six cells, while a 70 ampere hour one would usually have 11 to 13 plates per cell, with six cells.

When the plate assembly with its separators is assembled in its cell, it is connected in series with the next assembly by welding lead bridging members over the appropriate plate assembly vertical projections, or posts—one of which is shown, sectionally, on the lower left hand side, in Fig. 149. The cells are provided with moulded hard rubber covers, which allow two acid seals for the two terminal posts and a central combined filler and gas vent hole. These covers are fixed over the plates in their cells, before the end posts are connected by the bridging members; their edges are made liquid-tight by running melted pitch around the joints.

Later Improvements. The exposed terminal and bridging bar type of battery has the disadvantage of ready liability to short-circuiting of the plate terminals, when work is being done on the car, by mechanics' metal tools. Also, acid is liable to spill, when replenishing or 'overflowing' the cells with water, or by splashing, so that the metal around—including the steel holding down straps—are readily corroded.

FIG. 150. Improved, enclosed top cover battery

Later designs of battery completely enclose all the cell connectors leaving only the tapered lead alloy terminal posts exposed. Fig. 150 shows the Lucas SL7 battery with entirely enclosed intercell connectors and also a special cover which carries the six rubber vent plugs, which are a push-fit in the cell filler holes in the battery lid. A manifold venting system is provided to allow escape of the gases generated in the cells. The cover is lifted off bodily, to expose the six filler holes for topping up or density measurement purposes.

Various types of cell vent-filler plugs have been used to act as splash guards and, in some instances to ensure that the cells are filled or topped-up to their correct level, automatically.

Armoured Plate Batteries. This type of commercial vehicle battery has so-called armoured positive plates in which the active material is enclosed in shrunk-on, perforated ebonite tubes or, alternatively, the thicker-type positive plates are enclosed in perforated ebonite envelopes or are protected by sheets of glass wool. Such batteries withstand severe service conditions over longer periods than those of the normal positive plate construction. The C.A.V. armoured plate battery has 'fingers' completely armoured with shrunk-on perforated ebonite and with ebonite protection of the lower bar. The separators are of chemically-

THE AUTOMOBILE BATTERY 239

treated wood, and there is a porcelain part at the bottom of the strong ebonite case, acting as a rest for the plates and providing a sediment space below.

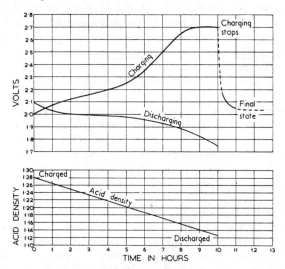

FIG. 151. Battery charge and discharge voltage characteristics

Battery Performance. Fig. 151 illustrates the voltage changes when a cell is charged at what is known as the 10-*hour* rate and also when it is discharged for 10 hours, so that the electrolyte density falls from the fully-charged value of 1·28 to about 1·125. During charging the voltage of the cell rises continuously until at about 6 to 7 hours there is a more rapid increase in the cell voltage, to attain its maximum value of 2·7 volts at 9 to 10 hours. When the charging current is switched off the cell voltage falls rapidly to about 2·25 volts and thereafter finally settles down to about 2·0 to 2·1 volts. The discharge graph characteristics depend upon the rate of discharge and the permissible final acid density. As shown in the lower graph (Fig. 151) the cell is discharged from its initial fully charged condition, over a period of 10 hours, when the discharged condition, corresponding to an acid density of 1·125 is attained. If the cell is permitted to become fully discharged and left for a period there is a risk of formation of insoluble lead sulphate, with possible distortion of the cell plates.

Capacity of Battery. The capacity or rating of a battery is a measure of the energy stored in it and is expressed as the period during which the battery will give the rated current before reaching the specified final voltage. It is usual to state the capacity of a battery as the product of the current, in amperes, and the time, in hours, before the battery voltage falls to 1·8 per cell.

Battery Ratings. If a battery discharges over a period of 10 hours with a constant current of 5 amps. it is said to have a capacity of 5 × 10, or 50 ampere hours at the *10-hour rate* of discharge. Most car batteries are rated on a *20-hour* discharge period, so that a battery of, say, 65 amp. hrs. would give a current of 3·5 amps. for a period of 20 hours, before the cell voltage fell to 1·8.

The 20-hour rating gives a measure of the lighting capacity of the battery. There is also a short-time rating which gives an indication of the ability of the battery to supply current for engine motor starting purposes; in this connection, as shown elsewhere in this volume, the starting current for a cold engine is of the order of some hundreds of amperes, for the short starting period.

In the U.S.A. a typical *short time rating* is the number of minutes that the battery will deliver a current of 300 amperes at a starting temperature of 0° F (−18° C). When testing a battery with this discharge current, the terminal voltage of a 6-volt battery should not fall below 3 volts and for a 12-volt battery, 6 volts.

For heavy vehicles such as motor coaches or buses, the 4-hour rating is often used. The U.S. specification is for the battery to be brought to 80° F and discharged at a rate equal to one-quarter of the published 4-hour rating in ampere hours, until the voltage has dropped to an average of 1·75 volts per cell.

For automobiles, with the electrolyte at 80° F the battery is discharged at a rate equal to $\frac{1}{20}$ of the stated 20-hour capacity, in ampere hours, until the terminal voltage is 1·75 volts per cell,

THE AUTOMOBILE BATTERY   241

i.e. 5·25 volts for a 6-volt battery and 10·5 volts for a 12-volt one.

Battery Capacity and Discharge Rate. The manner in which the capacity of a battery is affected by the rate at which it is discharged, and also the terminal voltage per cell at each discharge period are shown in Fig. 152, for a 50-amp. hr. cell, rated to give 2·5 amps. for 20 hours. The capacity ordinates denote the capacities as percentages of that of the fully charged cell. It will be seen that the 10-hour rating gives 90 per cent of the 20-hour

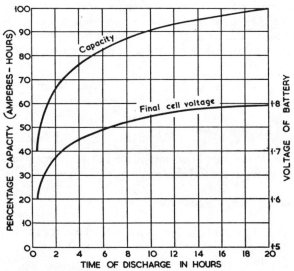

FIG. 152. Showing discharge rate effect on capacity of battery

rating, i.e. a current of 4·5 amps. for 10 hours. For a 4-hour rate of discharge the capacity drops to about 76 per cent of the 20-hour rate value.

Typical Battery Capacities. For small cars, up to 1 litre engine capacity, the batteries used range from about 30 to 40 amp. hrs. at the 20-hour rating. For $1\frac{1}{2}$ litre to 2 litre cars the capacities are from 43 to 50, and for 2 to 3 litre cars, 50 to 70 amp. hrs.—all at the 20-hour rating.

Typical American cars with engines of 325 to 400 b.h.p. outputs have batteries of 70–75 amp. hrs. The majority of the other mass produced cars with engines of 200 to 300 b.h.p. outputs and the smaller and 'compact' cars use batteries of about 45 to 55 amp. hr. capacities.

Batteries used in the heavier commercial and passenger vehicles have capacities, at the 10-hour rate of 90 to 240 amp. hrs. in 6-volt units. Four such units are joined in series to provide the standard 24-volt supply, in this country.

Passenger vehicles require the higher capacities, namely 140 to 240 amp. hrs. in the larger vehicles, extending in special cases up to 300 amp. hrs., when Diesel engines are employed—since these engines require an appreciably higher motor starter current.

Battery Efficiency. It is usual to define the efficiency of a battery in terms of watt-hours or ampere hours.

Thus Efficiency (watt hour) $= \dfrac{W^1}{W^2}$

where $W^1 =$ watt hour output for full discharge and $W^2 =$ watt-hour input for full charge

Efficiency (ampere hour) $= \dfrac{W^3}{W^4}$

where $W^3 =$ ampere hours for full discharge and $W^4 =$ ampere hours for full charge, from discharged state.

Effect of Temperature on Battery Characteristics. It is well known that a car battery is less active when very cold than at normal air temperatures. This is due to the effects of cold on the electrolyte since, as the temperature falls its electrical resistance increases. Its density also increases so that the liquid cannot diffuse so well within the pores of the battery plate material. On both accounts the capacity of the battery falls with a decrease in its temperature.

In regard to engine cold starting, the lower capacity of the battery is accompanied by increased viscosity of the cylinder and bearings lubricating oil so that a greater starting effort or torque must be supplied by the starting motor. It has been recorded* that if the power required to crank an engine at 80° F (26·7° C) with S.A.E. 20 oil is denoted by 100 per cent, at 32° F (0° C) the power required is 155 per cent and at 0° F (−17·8° C) it is 250 per cent.

In regard to the battery's energy, expressed as cranking power, the effect of temperature at three different states of charge, namely, Full, Half and Almost Discharged, is shown, for temperatures of 80° F, 32° F and 0° F in the accompanying Table, in which the results are expressed in terms of a fully-charged battery at 80° F, namely, 100 per cent cranking power; the electrolyte density is given as 1·280.

Relative Cranking Powers at Different Temperatures

Cranking Power 80° F	Full Charge	Half Charge	Almost Discharged
	100	46	25
Acid Density	1·280	1·225	1·180
Cranking power 32° F	Full Charge	Half Charge	Almost Discharged
	65	32	16
Acid Density	1·280	1·225	1·180
Cranking Power 0° F	Full Charge	Half Charge	Almost Disaharged
	40	21	9
Acid Density	1·280	1·225	1·180†

† Battery in danger of damage by freezing.   (Auto-Lite Corpn.)

Effect of Temperature on Electrolyte Density. The electrolyte used for lead-acid type batteries consists of a solution of one volume of brimstone sulphuric acid (of density 1·835) to 2·9 volumes of distilled water; this will give an electrolyte density of 1·280 at 60° F (15·6° C). Some manufacturers of batteries recommend other densities for their products, namely, from about 1·275 to 1·320.

* Auto-Lite Corporation (U.S.A.)

It is usual to stipulate a standard temperature of either 60° F (15·6° C) or 80° F (26·7° C) at which electrolyte densities should be measured. If the temperature of the electrolyte is above or below these standards, the readings should be corrected to 60° F or 80° F, as follows:

If the temperature is above 60° F or 80° F add ·004 for every 10° F in order to obtain the corrected reading. Thus, if the temperature is 85° F, then ·010 should be added for the 60° F standard, and ·002 for the 80° F standard density, e.g. 1·280. The corrected densities for 60° F and 80° F standards (1·280) will therefore be 1·290 and 1·282, respectively.

If the temperature at which the reading is taken is below either of the standard temperatures, then ·004 should be *deducted* from the density reading.

Measuring the Density. Since it is agreed that the acid density gives a more reliable indication of the state of a battery it is usual to measure the density with an hydrometer. Special types of hydrometer used for automobile batteries are supplied

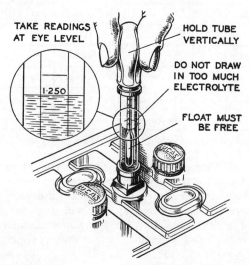

Fig. 153. Car battery type hydrometer, used also for topping-up purposes

THE AUTOMOBILE BATTERY

with a rubber bulb and a flexible suction tube so that when the cell vent or filler is unscrewed, the acid can be sucked up into a glass tube which contains a float-type hydrometer. The depth to which this float rises or sinks in the acid is shown by an internal scale in the vertical tube part of the float; the scale is graduated in acid density divisions.

Fig. 153 shows a typical car battery hydrometer with its suction tube (not shown) inserted into the top of the electrolyte. The liquid has been drawn up into the outer glass tube, so that the density reading can be taken; instructions for taking correct readings are given on this illustration. It is, of course, necessary to read the density of the acid in each individual cell, i.e. three cells for a 6-volt and six cells for a 12-volt battery; usually, the average of three or four readings should be used for the final reading.

Testing the Capacity of a Battery. The state of charge of a battery is often measured with an instrument which inserts a resistance across the cell terminals and measures the cell voltage reading. The current flow across the resistance is high, namely, 100 to 200 amperes; for this reason the tests are of very short duration, namely, a few seconds. Fig. 154 shows a typical battery testing unit.

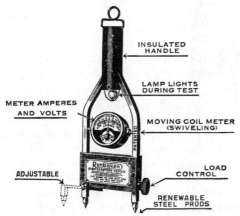

FIG. 154. Battery tester for load-current and voltage tests

The value of the resistance used is such that the cell voltage reads about 1 volt (depending on the battery capacity) for a fully-charged battery. A difference of more than 0·1 volt indicates a low cell charge.

A recommended test for a 12-volt battery is to connect a resistance of 0·08 ohm across the terminals, so that a current of about 150 amperes flows across. The voltage across the terminals is then read off. If each cell is fully charged the battery voltage should not fall below 10 volts; further, all cells of the battery should give the same reading. Lower voltage readings indicate faulty cells or cells failing to hold their full charge.

The Cadmium Test. In this accurate method of testing the condition of a cell a cadmium rod is inserted in the cell acid, the rod being connected to one terminal of a central zero type of voltmeter, reading 0-3 volts on one scale and 0-0·3 volts on another scale. When the cadmium rod is in the acid a layer of cadmium sulphate is formed on its surface, which is neither electro-positive nor electro-negative to the acid in the cell, so that if voltmeter readings are taken between the cadmium electrode and either of the cell's plates these readings will give the potential due to the respective plates when charging takes place.

FIG. 155. Cadmium test meter, for battery condition testing

A fully charged cell gives a cadmium-to-plate reading of 2·35–2·50 on the positive plates and minus 0·1 to 0·2 on the negative plates; the correct voltage is the algebraic sum of the two readings, e.g., $2·50 + 0·1 = 2.60$ volts.

A charged cell should give a positive reading of 2·35 volts and a discharged cell, not less than 2·00 volts, while the negative plates should not be less than $-0·10$ volt (charged) and not greater than $+0·25$ volt (reversed polarity) when discharged.

Fig. 155 shows the Ferranti cadmium test meter with its

special voltmeter and electrodes. The cadmium test serves as a final diagnosis in cases where a cell is known to be faulty, but where neither the acid density nor voltage test reveals the cause.

Cycling Effect on Capacity and Battery Life. The term 'cycling' as applied to the battery refers to repeated charging of the battery fully and then discharging at a normal rate until a voltage of 1·8 volts per cell is reached. The acid density and voltage of each cell should be taken at 30-min. intervals. The ampere-hour output then denotes the battery discharge, while any faulty cell reaches its voltage earlier than a good cell. For a battery in good condition a slight increase—usually about 1 per cent—in the capacity is obtained for the first cycle, but this usually indicates that the second cycle discharge takes longer, before the specified discharged condition is reached. Any such increase in initial capacity is obtained at the expense of a reduction in the total life cycles of the battery.

Life Cycling Tests. The quality and performance of a battery is determined by some manufacturers by means of life tests, consisting of charging and discharging at 1-hour rate, continuing these tests until the battery capacity is reduced to 30 to 40 per cent of its initial 20-hour rating. The total number of cycles before this capacity is reached is taken as a measure of the battery's status.

The Society of Automotive Engineers Specification for cycling tests of storage batteries (1961) is intended principally to evaluate battery internal components, and does not therefore attempt to simulate battery performance in the automobile.

In the case of 12-volt batteries below 80 amp. hrs. it specifies that, in all sizes, the battery shall be discharged for 1 hour at approximately 20 amp. for a total of 20 amp. hr. For 6-volt batteries these figures are doubled. The battery is then recharged at about 5 amp. for 5 hours for a total of 25 amp. hr. For 6-volt batteries the current and ampere hours are doubled. The time for one complete cycle is 6 hours. The average temperature of the electrolyte in the centre cell is kept at 105°–115° F.

To determine the battery condition during a life test, one

complete capacity discharge test is made on the fully-charged battery at 20 amperes (40 for a 6-volt battery) each week.

The length of time, in hours, for the battery voltage to fall to 1·70 volt per cell, multiplied by 20 (or 40) gives the ampere hour capacity at this rate.

When the capacity on a complete discharge cycle falls below 40 per cent of the manufacturer's original 20-hr. rating, the tests are terminated.

A new battery completing these cycle tests is required to withstand a certain minimum number of cycles, depending upon the battery size. Usually 100 cycles is considered to be satisfactory, but 200 cycles have sometimes been attained.

Battery Cycling, in Practice. It has been mentioned earlier that the discharge rate can affect a battery's capacity, if the discharge current is too high. Similarly, if with the normal discharge rates the battery is repeatedly charged and discharged, not only will the capacity in time be reduced, as stated, but the battery plates deteriorate due to *loss of the paste* and also, *grid corrosion*; the former fault is caused by volume changes of the plate material, during the discharge process, due to the lead sulphate conversion. Cars which are much used at night—more particularly in cold weather—usually discharge their batteries and re-charge when the load eases—as when parking with the lights on. In this connection the adjustment of the voltage, or current-and-voltage regulator, cannot altogether prevent these cycling effects.

Internal Resistance of Battery. When considering battery circuits, it is necessary to take into account the resistance of the battery itself, i.e. the plates and electrolyte. In this connection, according to the size of the plates, state of charge or electrolyte density and temperature, the resistance of the battery has the low value of 0·01 to ·001 ohm per cell. If the density is higher than the normal value, the resistance also will be higher. For the minimum internal resistance the acid density should be kept between 1·15 and 1·30; the actual minimum resistance density is 1·22.

Due to its internal resistance, when a cell is discharged the electromotive force is greater than the terminal voltage, and when charged is less than the applied voltage. It follows from this that the potential difference, or voltage V, between the cell terminals when *on charge* will be given by: $V = E + RC$ where $E =$ electromotive force of the cell, $R =$ internal resistance (ohms) and $C =$ charging current (amperes).

When the cell is being *discharged*:

$$V = E - RC$$

In the case of a multi-cell battery, with six cells the voltage at the terminals is 12 V. The internal resistance is not, however, constant. When fully charged the internal resistance is a minimum, but when fully discharged the resistance may be from four to five times the charged value.

Battery Charging. Each type of battery should be re-charged at its recommended normal charging current and period, as indicated in the manufacturer's instructions. When the battery is in a low charge state it will require a greater charging current than when it is nearly at full charge, so that for automobile purposes the charging current is adjusted to suit the state of the battery at all times by means of the dynamo regulator—as described in Chapter 3. Since it is possible to adjust a regulator to give a greater or lesser range of charging currents, care must be exercised to prevent *overcharging* of a battery, otherwise the battery plates may be damaged.

Another important factor is the engine (or dynamo) speed at which the cut-out acts, i.e. connects the battery to the dynamo. This speed should be low enough to keep the battery charged under all driving conditions, except at the dynamo speeds corresponding to low road speeds—usually, below about 12–15 m.p.h.—on top gear.

When an automobile battery is removed from the car, if the latter is out of action, or laid up, it should be kept charged by means of a battery charger, operating off the electricity mains, or any other independent source of electricity supply.

Battery chargers should incorporate a rectifier for conversion of

A.C. to D.C. supply; a rheostat to regulate the charging current and an accurate ammeter and voltmeter; the latter to check the charging voltage and voltages after the current is switched off.

If more than one battery is to be charged the *Constant Current* method of charging can be employed. In this case the batteries are connected in series, thus leaving a free positive terminal on one end battery and a free negative terminal on the other end.

It is necessary only to connect the battery charger positive supply lead to the free battery positive terminal and the negative supply lead to the free negative terminal.

Another method, known as *Constant Potential* charging consists in supplying current at a constant voltage from the battery charger and connecting the batteries in shunt between the constant voltage leads, with the positive and negative terminals of the batteries connected to the charger positive and negative, respectively.

In this case the initial charging rate or current is relatively high but, as the charging proceeds the charging rate tapers off to a lower value, depending upon the condition of the battery and the electrolyte temperature.

In the constant potential two-lead system only batteries of the same terminal voltage can be charged together and the charging currents for each battery will depend upon the condition and internal resistance. It is possible, however, to charge both 6- and 12-volt batteries at the same time by using *three leads or busbars*, namely a central or neutral conductor and two outer ones. A constant 7·5 volt supply is maintained between each outer conductor and the neutral conductor, so that the 12-volt batteries can be connected across the outer conductors (15 volts) and the 6-volt batteries across the 7·5 volt outer and neutral conductor.

High Rate Battery Charging. More recently a fast charging type of battery charger has been developed for boosting up the charging process, so that by employing currents of 40 to 100 amps., depending upon the size of the battery, it is possible to recharge a battery to nearly full charge condition in one hour if certain precautions are observed, namely:

(1) The battery must be in fully satisfactory condition, otherwise the plates may suffer serious damage.
(2) The electrolyte temperature must not be allowed to rise above 125° F.
(3) The battery, after the initial boosting charge—which should taper down to about 30 per cent—should be given a completion charge at a normal charging rate.

The fast charge time for a typical 60-amp. hr. car battery would be one hour, for an electrolyte density of 1·150, using a 40-amp. current. The charging time, however, will be less for batteries starting with higher electrolyte densities. Thus, if the density is 1·175–1·200, the charging time is reduced to $\frac{1}{2}$ hour and if 1·200–1·225 to $\frac{1}{4}$ hour only. When the density reaches 1·225 the battery should be given a finish charge of 4–6 amps., so as to bring the density up to 1·280.

High rate chargers should use the constant potential method and, after the initial high charging period should switch, automatically, to the slow rate finishing method; they should, preferably, be fitted with an electrolyte temperature limiting switch device to prevent temperatures exceeding 125° F.

The number of high rate charges that can be given to any battery must be strictly limited, namely, to 5–10 times, during its life; otherwise the battery life will be shortened appreciably.

In general, the high rate charger, if used according to the manufacturer's instructions on good condition batteries only, will give satisfactory results. The charger should not, however, be used on new batteries, made 'wet' from their original dry state.

Battery Plate Sulphation. Under normal operating conditions lead sulphate in a finely divided state will be formed during the cell discharge and become readily changed into lead and lead oxide, during recharging. If, however, a battery is left in the discharged condition for any appreciable time, i.e. if the acid density is allowed to fall below 1·15 per cell, or the terminal voltage below about 1·8 volts, then the lead sulphate is converted into its basic form, namely, a hard white non-conducting and crystalline substance.

The positive plate then changes from a chocolate brown to a pale or yellowish brown colour, while the negative plate alters from its 'leady' to an almost white colour.

As the sulphate deposits are now non-conducting, the internal resistance of the cell is increased appreciably, so that voltage and capacity are reduced, on discharge, while the cell may require about 2–3 volts more to pass the normal charging current through the cell. As the hard sulphate is of greater volume there is a tendency to buckle the plates.

It is sometimes possible with batteries not badly sulphated to restore their plates to approximately their normal condition by prolonged charging at low current rates, in a much weaker acid solution than that normally used for the electrolyte. When the acid reaches a density of 1·150, it should be replaced by distilled water.

Some Other Battery Faults. *Overcharging* of a battery, i.e. prolonging the charging time necessary to restore the electrolyte density to about 1·280, produces excessive heat which will cause the positive plates to expand and buckle or warp. Prolonged overcharging can also cause *oxidation* of the *plate grid metal*, with eventual breakdown of the plate; this is due to the excess nascent oxygen, produced during charging, attacking the grid metal, which becomes reduced in section.

Corrosion of the battery terminals, due to leakage of the acid past the battery top sealing material, resulting in the formation of white lead sulphate, is another more frequent trouble. To avoid this, the top of the battery must be kept dry and any observed place of leakage sealed with a suitable filler or washer. Finally, the complete battery clip should be liberally coated with vaseline or non-acid grease, after it has been secured to the battery post; the latter and the inside of the clip must, of course, be quite clean.

Loss of charge over a period after a battery has been fully charged can be due to excessive deposits in the bottom of the battery moulded container, caused by loss of active material from the plates. This deposit, if severe, can actually bridge the two sets plates of a cell and thus short-circuit them.

Worn, pitted or otherwise *damaged plate separators* can lead to loss of charge and reduced battery capacity.

Dry Charge Batteries. When supplied in the *dry state* the positive and negative plates are in the fully charged condition, but there is no electrolyte in the battery. This type of battery if stored in a dry cool place, can be kept for at least three years without deterioration.

When acid of the correct density, e.g. 1·270 at 80° F* is poured into the cells, after breaking the hermetic seals, it is soon ready for service, but it is advisable to allow the battery to stand for 12 hours, to allow the electrolyte to soak into the plates and wood separators (when used). If the acid density is correct the battery can be used without further charging, but it is usually advisable to give it a low charging rate for at least 12 hours, after which the acid density should be checked.

In the case of *new uncharged batteries*, after the seals are broken, acid solution of the density recommended by the manufacturers is poured slowly into each cell to the correct level. The battery should then be allowed to stand for at least 12 hours.

In connection with the initial filling of the new battery, since new cells can soak up to about 20 per cent of the acid, the cells will require topping up with acid, after the 12-hour standing period.

The first charge should be made at the recommended current rate; this is usually *two-thirds* of the subsequent *recharging rate*. The charging period at the initial current value usually varies between 30 and 100 hours (for commercial sizes). The charging process can be stopped when the cell voltage is 2·55 with the charging current flowing and when after three successive checks there is no further voltage rise. The acid density should attain the manufacturer's value, after three successive checks.

As an example, a 63-amp. hr. battery, having 9 plates per cell, would be filled with acid of 1·275 density and given an initial charge at 4·5 amps. until, as explained earlier, it is fully charged, when the acid density should be 1·280–1·300. For tropical

* 1·320 at 60° F for some commercial batteries, giving 1·280–1·300 fully-charged density.

temperatures the initial and final acid densities would be 1·260 and 1·220–1·240.

After the initial charge, the battery when partly discharged should be recharged at 7 amps.

The Alkaline-type Battery. It has long been acknowledged that the lead-acid battery is by no means the ideal for automobile purposes. Its chief disadvantages lie in its relatively heavy weight for a given capacity; its liability to sulphate formation if left partly or fully discharged; its self-discharge action if left standing for appreciable periods; its sensitiveness to shock and vibration; its liability to damage if overcharged, or if discharged above a certain current-time rate and its unsuitability for use at lower temperatures, i.e. to freezing up, more especially at low acid densities.

Before describing the alkaline cell, it is proposed to enumerate briefly its principal advantages. The alkaline cell:

(1) is lighter in weight than the lead-acid battery of equal capacity;
(2) is much stronger, mechanically, and will withstand severe shocks and vibration without detrimental effects on the plates or case;
(3) has a much longer useful life, namely $2\frac{1}{2}$ to 4 times that of the lead-acid battery;
(4) is not damaged by overcharging or rapid discharging;
(5) can be left for long periods discharged, without deterioration;
(6) has extremely low self-discharge, if left charged over a long period;
(7) requires considerably less maintenance attention; this cost saving to some extent compensates for the higher initial cost of the battery;
(8) is able to withstand low atmospheric temperatures without damage.

Types of Alkaline Cell. There are two kinds of cell, namely, the nickel-iron and nickel-cadmium.* In each type the positive plate employs nickel peroxide material and the electrolyte

* Vide Fig. 158.

consists of a solution of caustic potash (potassium hydroxide) in water. The normal electrolyte density is 1·17 to 1·19 and this remains constant during the charging and discharging period.

Action of Nickel-Iron Cell. The negative plate of this cell is of pyrophoric active material held in steel tubes or pockets having a large number of minute holes, the tubes or pockets being held in steel retaining plates; these cannot buckle and the material is not dislodged by vibration.

When the cell is discharged the nickel peroxide $Ni(HO)_2$ is reduced to nickel oxide NiO while the iron on the negative plate is converted to iron oxide FeO.

When charging, the iron oxide and nickel oxide are changed to pure iron and another nickel oxide, Ni_2O_3. Actually, the chemical intermediate processes are rather more complex, but the final products are as previously stated.

During the charging process oxygen is liberated at the positive and hydrogen at the negative plate, so that provision is made for the escape of these gases, while suitable precautions regarding risk of inflammation of these gases must be taken.

Regarding the chemical changes which occur in charge and discharge operations the electrolyte acts mainly as a conductor and, as mentioned earlier, its density remains unchanged except for electrolysis. Therefore, the hydrometer density test as an indication of the state of charge cannot be employed with the alkaline cell.

With regard to the electrolyte, it must be protected from exposure to the air, otherwise contamination, due to carbon dioxide may occur.

During the charging process (Fig. 156) the cell voltage rises from about 1·5 volts at commencement, to 1·6 to 1·8 volts, according to the state of charge. When charging, the temperature of the cell and its contents must not exceed 110° F (43° C), otherwise the active materials of the plates may deteriorate.

These cells will withstand heavy charging rates, and also overcharging without damage; similarly, rapid discharging if not made a practice does no harm. The normal discharge rates should, however, be employed; the terminal voltage on discharge should

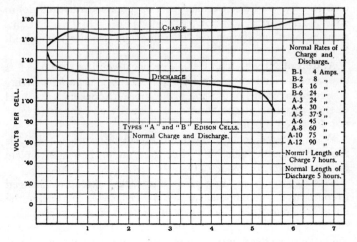

FIG. 156. Charge and discharge curves for nickel-iron battery

not be below 1·1 volts per cell. During a rapid discharge the voltage falls off at a lower rate than for a lead-acid cell.

Fig. 157 shows how the capacity of an alkaline cell is affected by the rate, or time of discharge. It will be seen that over the period of 3 to 10 hours the capacity does not fall by more than 4 per cent, so that for these discharge rates the capacity is practically constant.

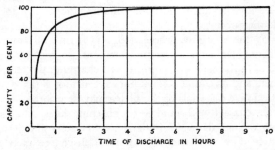

FIG. 157. Effect of rate of discharge on capacity of alkaline cell

THE AUTOMOBILE BATTERY

Temperature Effects. The effect of temperature rise from 68° F (20° C) to 122° F (50° C) is to increase both the capacity and the voltage; below 60° F the capacity and voltage of the cell falls.

Some Notes on Alkaline Batteries. In view of the completely different characteristics of this type of battery, the following brief information should be of interest.

Indications of State of Charge. It is more difficult to measure the state of charge, since the electrolyte density test is inapplicable to alkaline cells. The voltage test, also, is not reliable, but if voltage readings are taken with a given charge or discharge current, some idea of the battery charge can be obtained. It is usually recommended that the *ampere-hour meter* be used to ascertain the state of the charge, but it is necessary to correct for the variation of ampere-hour charging efficiency; this varies from about 90 per cent for cell voltages of 1·3–1·5 volts and to 50 per cent for 1·5–1·7 volts, at the 10-hour rate of discharge.

The Electrolyte. The electrolyte density is not as important as with the lead-acid cell since a variation of about 15 to 25 per cent does not affect the performance of the cell. It is necessary to top up the cells to their prescribed levels, with distilled water, to make up for loss of liquid due to escape of the gases generated.

It is recommended that the alkaline solution should be emptied out after about 300 discharges of the battery, the cells washed out with distilled water and fresh electrolyte introduced. It is unusual to find any deposits in alkaline cells. An indication of the necessity to replenish the electrolyte is when the density falls to about 1·16, regularly, after normal discharge.

Before emptying the electrolyte from the battery, it is important that the battery is completely discharged to a voltage of 1·1 per cell. The battery plates must not be exposed to the atmosphere for more than 20–30 minutes.

Charging. For the first charge in the case of a new battery it is usual to employ the recommended charging current, normally used for recharging, but to allow twice the normal charging period. It is necessary to increase the cell voltage from its initial charging value of 1·4 volts, as charging proceeds, in order to maintain the current constant. The actual charging rates depend upon the size and type of battery, so that the manufacturer's figures should be employed.

In general, standard battery chargers and charging methods for lead-acid batteries can be employed for alkaline ones.

Corrosion. Since the alkaline electrolyte has a preservative effect on steel parts used in the batteries, the battery terminals do not corrode, but it should be mentioned that the electrolyte (diluted caustic potash) can injure both the skin and clothing of individuals, so that suitable precautions must be taken. Boracic acid solution or 10 per cent citric acid in water, will alleviate the effects of skin burns.

Internal Resistance. With the usual constructional methods, alkaline batteries will have higher internal resistances than lead-acid ones. This will result in a larger difference between the minimum voltage on discharge and maximum voltage on charge. The wider difference renders engine starting more difficult, where large current values are needed and also, these voltage variations affect the life of the vehicle's lamps. When these drawbacks were fully realized, the manufacturers proceeded, by the use of new constructional methods, including the use of larger but thinner plates and special ingredients to reduce the internal resistance, so that the alkaline battery operated more favourably and became rather more widely adopted for commercial vehicles, motor cycles, etc.

The Nickel Cadmium Battery. This type of battery, which is represented by the Nife and Nife–C.A.V. types, is very similar in principle to the nickel iron battery, but while using the same alkaline electrolyte, employs cadmium instead of iron.

When a nickel cadmium cell is charged cadmium oxide is produced on the negative plate; this is reduced to spongy cadmium on discharge. Similarly to the nickel-iron cell the nickel oxide is converted to nickel peroxide on charge.

THE AUTOMOBILE BATTERY

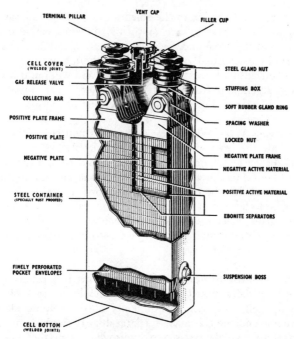

Fig. 158. Construction of C.A.V.-NIFE nickel cadmium battery

When a nickel cadmium cell is fully charged the charging voltage rises from 1·35 to 1·60 volts, which is rather lower than for the nickel iron cell which gives an increase from 1·6 to 1·8 volts. However, the open circuit voltage of each type is 1·25–1·4 volts.

The normal density of the electrolyte is 1·190 for the C.A.V. battery; if the density is higher than 1·200 or lower than 1·160 the cell will not operate satisfactorily.

The density of the electrolyte relates to a temperature of 68° F (20° C). At 48° F, 88° F and 108° F the corresponding densities are 1·195, 1·185 and 1·180, respectively. The electrolyte should be maintained at a level of $1\frac{1}{4}$ to $1\frac{1}{2}$ in. above the tops of the plates.

Fig. 158 illustrates the constructional features of the C.A.V. nickel cadmium battery which is made in a range of models for commercial vehicles from the smallest, of 45 amp. hrs. (at 10-hr. rate) to the largest of 200 amp. hrs. The respective charging currents at the recommended 7-hr. rate are 9 and 40 amps.

As supplied normally, the batteries, although not fully charged can be put into service on motor vehicles and will build up to the fully charged state from the dynamo; otherwise an initial charge of normal current for 7 hours will give the full charge.

Charging Notes. When the electrolyte is poured, to the correct level, into the cells, and the battery previously has been emptied in its discharged condition, the cells should be allowed to stand for 24 hours, to allow the electrolyte to soak into the plates. The cells should then be topped up to the correct level with distilled water and charged at the normal 7-hour rate, for 5 hours only. After again checking the electrolyte level, the cells should be charged at the normal 7-hour rate for 10 hours, after which the battery is ready for service.

The Venner Silver-Zinc Battery. This is a relatively light battery initially based upon the principle of the silver-zinc primary cell, but with the difficulty of the soluble electrodes overcome and the cell made rechargeable. It employs an alkaline electrolyte, namely potassium hydroxide, as in the nickel-iron and nickel cadmium cells and the action is similar in regard to the constant density during charge and recharge. The chemical reaction for the silver-zinc cell is:

$$AgO + Zn(HO)_2 \rightleftharpoons AgO + Zn + H_2O$$
$$\text{(Discharged)} \qquad \qquad \text{(Charged)}$$

The electrolyte is 'unspillable', due to the fact that it is absorbed.

It is stated that an 85-amp. hr. Venner battery of the 12-volt type weighs only 17 lb. It is much smaller and lighter than the equivalent lead acid or nickel-iron type but is considerably more expensive, namely, about 5 to 7 times that of the same capacity lead acid battery, after allowance is made for the 'scrap' value of the silver-zinc battery.

CHAPTER 9

THE LIGHTING SYSTEM

The lighting system considered in this chapter includes the headlamps, fog lamps, front and rear side (or parking) lamps, direction indicator or flashing lamps, brake and other warning lamps (fitted on instrument panel) and certain panel, interior illumination lamps with their individual switches. Occasionally a reversing lamp is fitted. The method of wiring all of these lamps is based upon the shunt principle, in which each lamp (with or without its own switch) may be regarded as being connected across the battery terminals, as shown schematically in Fig. 4, on page 21. Examples of complete wiring diagrams, showing the various lamps, etc., are given in Figs. 6 and 7, while fuller information regarding the wiring and installation of the lamps and various other items of electrical equipment is given in Chapter 12.

With the substitution of electric lighting on automobiles, for the previous oil and acetylene headlamps, the battery with its dynamo and regulator system was originally introduced, although the small 2-volt accumulator had previously been employed for operation of the trembler coil ignition system.

The modern lighting system incorporates the lamps, switches (individual or combined), fuses (or circuit-breakers in American practice) and the wiring harness.

Energy Demand of Lighting System

With the more widespread usage of electric bulbs and other equipment for various purposes, in modern automobiles, the battery current demand has increased considerably. In the average British car the lighting system is responsible for about 70 per cent of the total electrical energy when cars are driven at night. Similarly, for American cars, the percentage is also high, namely, about 72 to 75. In some instances, from 24 to 27 amperes may be required, at night, on British and 30 to 40 on American cars, for all purposes, including the heater, radio and transmission control.

The Headlamps. Modern headlamps are the results of research and developments, from the original type employing a single electric bulb of the carbon filament (and later vacuum-filled metal filament) type, placed at the focus of a parabolic silvered reflector so as to provide a parallel beam of light. In practice, owing to the fact that the lamp filament was not the assumed ideal optical point source of light, but a small cylinder or vee-bar, there was an appreciable amount of deviation from the ideal beam; also, there was a definite amount of light scatter from the front beam, which did not strike the reflector. By suitably proportioning the reflector, the amount of light scatter could be arranged to provide the necessary side illumination, nearer to the car. Fig. 159 illustrates a parabolic reflector with the light bulb at B, the arrowed lines showing light rays in all directions,

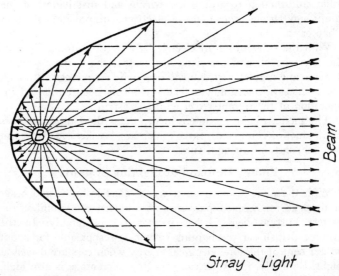

FIG. 159. The parabolic reflector, showing light rays from bulb at focus B

emitted from the bulb filament. The scattered or stray light rays are also shown in the diagram by the right-hand side divergent arrows. This form of headlamp gives a parallel beam of light which, with the exception of the small amount of light that is blocked by the bulb itself, gives greater illumination nearer the axis, with the light intensity falling off towards the outer part of the beam.

It can readily be shown that if the bulb filament is moved from the focus of the reflector *towards* reflector, the light beam fbf' will no longer be a parallel one, but will become divergent, as shown at *abc* in Fig. 160. If, on the other hand, the filament is

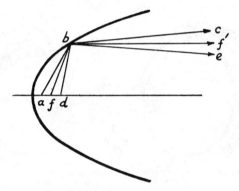

Fig. 160. Showing effect of altering the position of the lamp filament on the nature of the light beam

moved from the focus, away from the reflector, the beam will become divergent as shown at *dbe* and the light rays will meet at a point on the lamp axis. By suitable adjustment of the bulb position d the beam can be concentrated at a predetermined distance ahead, so as to give a *spotlight* effect.

The less recent headlamp bulbs were provided with means for adjusting the bulb holder in relation to the reflector, along the axis so as to focus the bulb; this was usually necessary, every time a bulb was changed. Unless the focusing of the filament was done accurately the lamp would have a poor range, and would cause increased dazzle to oncoming traffic.

Pre-focused Bulbs. The first major advance in headlamp design was the introduction of the headlamp bulb with its filament accurately positioned in relation to the bulb's metal seating flange. The reflector was also provided with means for locating the pre-focused bulb accurately at the focus of the reflector. This was effected by a slot in the bulb flange which engaged with a projection on the inside of the reflector's bulb-holder, which located the bulb correctly in the reflector. A bayonet fitting cap with spring contacts secured the bulb firmly in position and also carried the current supply to the bulb contacts.

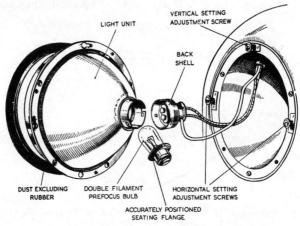

FIG. 161. Headlamp components for pre-focus lamp bulb

Fig. 161 shows the Lucas headlamp, fitted with the pre-focus bulb and reflector combination, on the centre and left hand side, and the car body mounting flange on the right. The light unit, consisting of the reflector, bulb holder and front cover glass forms a complete unit that is secured to the body mounting by a single screw, which, when removed, exposes three other spring-loaded screws. Three corresponding holes with slots, on the light unit flange allow the light unit to be placed over the heads of the adjusting screws and, by a slight rotation of the unit to lock the unit in position. The adjusting screws are also employed to tilt the light unit vertically—by the top screw—and horizontally by the two lower screws. The light unit is turned in an anticlockwise direction to remove it from the body mounting. The back shell (Fig. 161) is removed by moving anticlockwise and pulling it off.

The pre-focus bulb has two filaments, one for the normal driving beam and the other, which is operated by the driver's left side foot switch, for dipping the beam, when meeting oncoming traffic.

Sealed-Beam Headlamps. Introduced originally in the U.S.A., a special self-contained glass unit comprises the glass reflector sprayed inside with vaporized aluminium to give a reflecting surface comparable to that of silver. The unit front lens or cover is fused to the reflector, after the two-filament unit has been inserted through the centre of the reflector and sealed in position. The complete unit is then evacuated and filled with an inert gas (Fig. 162).

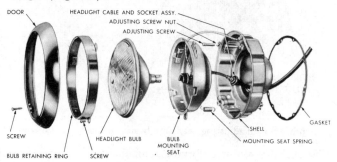

Fig. 162. American sealed-beam headlamp components

The advantages of the sealed-beam headlamp are (1) The glass unit is self-contained with accurately focused filaments. (2) All outside dust, moisture and other extraneous matter is permanently excluded from the back of the lens and the reflector. (3) Due to the absence of a filament bulb, a greater amount of light is provided in the beam. (4) Greatly improved beam, due to pre-focused filament and permanently bright reflector. It is necessary to provide for the vertical and horizontal adjustments of the beam, by tilting the sealed-beam unit in its body housing. This is usually effected in American cars by two adjusting screws, namely, a top central screw for raising or lowering the beam and a left-hand side screw for sideways, or horizontal adjustment of the beam, as shown in Fig. 163.

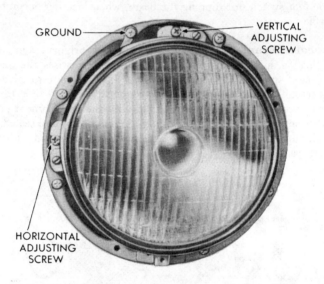

FIG. 163. Method of adjusting headlamp beam, vertically and horizontally

The sealed-beam or bulb unit is mounted in a pressed metal unit, having a central electrical contact connector to the sealed-beam protruding connections. This bulb-mounting is held in a fixed body mounting shell and the front of the bulb unit is secured by a bulb retaining ring (Fig. 162) over which is fitted a front cover, or door. The bulb unit is readily removed by first taking off the door and unhocking a spring from the retaining ring, when the bulb unit can be taken out after separating the unit from the cable connector at the rear.

More recent improvements to the sealed-beam unit include the provision of a metal mask, or shield in front of the upper filament

—which gives the lower beam—to prevent the stray light rays from escaping upwards, and reflecting back into the driver's eyes. Another improvement is that of providing a *reference plane* on the outside of the lens member, near the periphery. This consists of three equally-spaced flat surfaces or projections, such that their common plane is at right angles to the lens axis. Known as *aiming lugs* and with their outer faces precision ground they provide a reference plane for contacting the headlamp beam setting apparatus, now used for the initial setting of the light beams, correctly, on new cars and for subsequent rechecking when the sealed-beam unit is removed and replaced.

Fig. 164. Sealed-beam unit, showing aiming lugs

Fig. 164 shows the aiming lugs on the Buick headlamp sealed-beam unit; also the fluted lens pattern for horizontal beam spread and also depression of the upper part of the beams, to diminish or eliminate dazzle effects.

268 MODERN ELECTRICAL EQUIPMENT

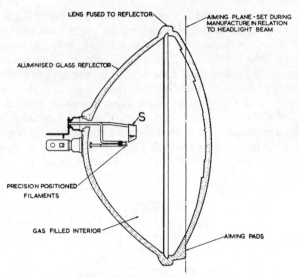

Fig. 165. The Lucas sealed-beam unit

The more recent Lucas glass sealed-beam unit, shown in Fig. 165 includes the later modifications, previously mentioned. The two filaments with the upper filament stray light shield (S), the filament outside connecting contacts, the internal fluting and the aiming pad plane, with one of the aiming pads, are clearly shown in this diagram.

In general it has been shown that the sealed-beam type headlamp gives a brighter beam pattern and reduces maintenance and repair time, although it is appreciably more expensive to replace such units than plain or pre-focused bulbs.

The light output of recent sealed-beam headlamps has been increased to 50 and 60 watts for the lower filament (upper beam) and 40–45 watts for the upper filament (lower or dipped beam).

Vertical and Side Control of Headlamp Beam. It was shown earlier that the single central filament lamp bulb, mounted at the focus of a parabolic reflector gave an approximately circular section parallel beam of light having the greater light intensity

THE LIGHTING SYSTEM

near to the axis of the reflector. From the viewpoint of the road user the circular beam has certain drawbacks. Firstly, such a beam does not provide sufficient side illumination to cover the width of the road, nearer to the car. Secondly, the upper part of the beam is distinctly of a dazzling nature to oncoming vehicles' drivers and, thirdly, much of the upper part of the beam is wasted by illumination of the unwanted part (above the approaching vehicle's top level) of the road scene.

The circular beam can be spread horizontally to any desired extent by means of refracting prisms moulded on the inside of the lamp front cover glass, or lens, as shown in Fig. 166 (A). By suitable design of such prisms or flutings the horizontal light intensity can be controlled in any desired manner.

The circular beam can be controlled in the vertical direction by means of similar prisms or flutings in the front cover glass, as

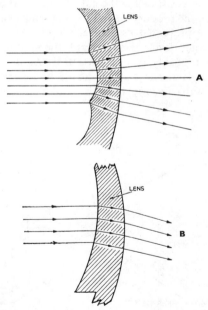

FIG. 166. Methods of altering lamp beams. (A) spreading beam horizontally. (B) deflecting beam vertically downwards

indicated in Fig. 166 (B), so that the dazzling upper part of the beam can be deflected downwards, below the eye level of the other road users. With a combination of vertical and horizontal flutings it is possible to produce any required main beam illumination, combined with a regulated amount of side illumination nearer to the car.

Non-Dazzle Light Beams. The problem of providing adequate illumination of the road ahead for moderate fast driving, combined with non-dazzling effects to other oncoming road traffic has been apparent since about 1918 and various suggestions for optical solutions of the problem have been devised since then. While it has been shown possible by optical means to produce what must be regarded as almost the ideal shape of driving beam, namely, one having a flat horizontal cut-off, with maximum intensity of the beam below the cut-off and decreasing intensity on either side, the production and costs of lamps based upon this principle have prevented its general adoption.

The first major development towards the solution of the dazzle problem was probably that of the anti-dazzle bulb patented by Graves in England and, later, adopted in European countries. The Graves bulb (Fig. 167) employed two filaments—an arrange-

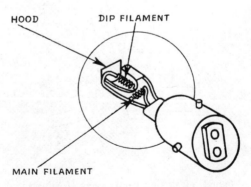

FIG. 167. The Graves two-filament light bulb

ment which has since become standard practice—one for the main driving beam arranged horizontally at the rear of another axially located filament. In order to provide a beam having a sharp horizontal cut-off a hood, or shield, was placed below the axial filament. When suitably located in its parabolic reflector, the latter while concentrating light from the filament, projected an image of the shield surrounding the filament, which resulted in a light beam having a flat top and a semi-circular boundary below, as shown in Fig. 168. Having produced this approximation

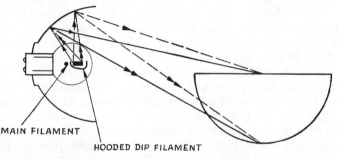

Fig. 168. The 'dipped' beam of the Graves bulb and reflector

to the ideal requirement, the semi-circular beam was given the necessary spread by means of the fluted front lens mentioned previously. The main or driving filament was located at the focus of the reflector and gave a parallel-type of driving beam. When the main beam was switched off and the auxiliary (axial) filament switched on, the screen cut off all the light rays that would otherwise have been reflected upwards by the lower part of the reflector, leaving the upper part of the reflector to produce the downwardly projected non-dazzle beam; this type of beam had a certain lateral spread which illuminated the road nearer to the car.

The chief disadvantage of the Graves bulb was that extreme accuracy was required in its production and also in the initial mounting of the bulb and its reflector in the headlamp, since any slight error in the location of the axial filament, relatively to its shield had a marked effect upon the shape of the beam produced;

further, it reduced the candle-power of the beam and thus restricted the driver's vision ahead to about 100 feet. Moreover, it gave equal illumination to both near and off sides of the road, whereas, a greater illumination to the near side is desirable.

The Dip-and-Switch Headlamp. This type of headlamp had a successful period of usage following that of the Graves or Duplo (or Dutch version) headlamp system. In the later type two headlamps with similar beams were used but, on meeting distant approaching traffic, a pedal-operated switch was actuated to switch off the off side (R.H.) lamp and at the same time to deflect the nearside beam downwards and to the left hand side so as to give a non-dazzle beam with increased illumination of the left side of the road.

The deflection of the beam was effected by an electromagnet in the nearside lamp cover, which was actuated by the pedal switch. As the pull needed to move the reflector was greater than that required to hold it in the tilted position the solenoid plunger, on reaching the end of its travel opened a pair of electrical contacts and inserted a resistance into the circuit, which reduced the solenoid current to a low value. When the current was switched off the reflector was returned to its original position by a spring. The side deflection accompanying the dipping of the reflector was about $5°$ to $6°$; this resulted in the beam meeting the side of the road about 100 to 120 ft. ahead of the car.

A simple but effective method, using ordinary headlamps is to fit the fixed offside headlamp with a rather higher wattage bulb than that normally used, and the nearside headlamp with the standard bulb. The nearside lamp is, however, focused so that its beam tilts downwards and to the left—as with the dip-and-switch lamp, but arranged for rather more forward road side illumination. The offside lamp is switched off when meeting approaching traffic, leaving the nearside non-dazzle lamp in action.

The Two Filament Modern Headlamp. The two filament bulb is used with a fixed reflector, in this case, with one filament for the dipping or meeting beam above, and the main beam filament beneath it and located at the focus of the parabolic reflector

THE LIGHTING SYSTEM

as shown in Fig. 165. A foot-operated switch is used to switch off the main filament current and switch on the upper filament. It will be seen from the diagram that in this way the upper filament, used in conjunction with a prismatic lens or light shield gives the beam a downward deflection. The nature of the dipped or meeting beam is controlled also by the prisms and lenses in the front glass. The nett result of this method—known as the Anglo-American system—is to give a high intensity driving beam and a rather lower intensity dipped beam just below the horizontal level of the lamp centres; at the same time, there is a good spread of light across the road and a high intensity below the horizontal plane, and towards the left or near side of the road.

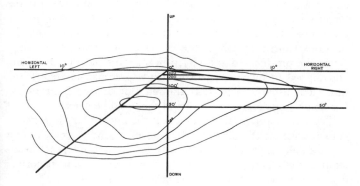

FIG. 169. Light intensity graphs for British dipped beam system

Fig 169 shows the light intensity graphs of the British dipped beam system in relation to the horizontal line ahead. It will be seen, from those parts of the graphs that are above the horizontal line, that there is only a small amount of light or glare seen by an eye at this level and that the maximum illumination, shown by the smallest graph, is well below the horizontal line and to the left of the road centre. The intensities of the light at different distances from the lamps up to about 400 ft. are shown on the centre 'Up-Down' line.

The maximum illumination, in the standard specification tests, corresponding to the smallest graph, was 8,000 candle-power.

Lamp Wattages. English pre-focused bulbs usually employ 45 to 50 watt main or driving beam filaments, and 30–40 watt dipped or meeting beam filaments. There is, however, a recent tendency to increase the wattages of the filaments by about 15 to 20 per cent.

The American (S.A.E.) specification graphs give somewhat similar characteristics, except that below the horizontal line the total width of the beam is greater than in the British specification; above the horizontal line, on the driver's off-side, there is a similar relief from dazzle in the American specification.

Aligning Headlamps. It is important that the headlamps on cars and other motor vehicles are aligned accurately, in relation to the longitudinal symmetrical axis of the car, so that the driving beams are parallel to this axis and also at the correct height above the horizontal plane on which the car is assumed to be standing. Since the pre-focused and sealed-beam units have their driving and dipped or meeting beam filaments precision-located in the units it is usual to align only the main or driving beam of the headlamp, the dipped beam should then also be correct.

The general procedure for practical alignment of the headlamps is to arrange for the car to stand on level ground at 25 ft. from a vertical wall, preferably white, or with a white screen for each headlamp, mounted on the wall. The car should be carrying its normal load and its tyres inflated to the correct pressures.

The covers of the headlamps should be removed (Fig. 170) and the vertical and horizontal adjusting screws turned until each beam is parallel to the longitudinal axis of the car. The centres of each bright area of concentrated light should be separated by the distance between the centres of the lamps on the car and the height of the centre of the bright areas should be the same as the height of the centre of each lamp above the ground, as indicated for the Lucas-type headlamps, in Fig. 171. It is usual to cover one headlamp while the other is being adjusted. If two vertical chalk marks are made on the wall at the same distance apart as the centres of the headlamps, and a horizontal chalk line is drawn at the height of the centres of the lamp—as indicated by the dotted lines in Fig. 171, these lines will simplify the alignment procedure.

THE LIGHTING SYSTEM

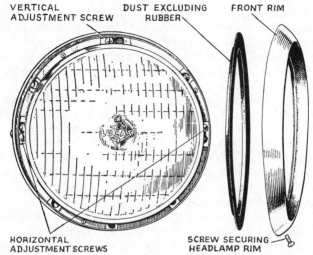

FIG. 170. Horizontal and vertical adjustment of Lucas headlamp (Standard)

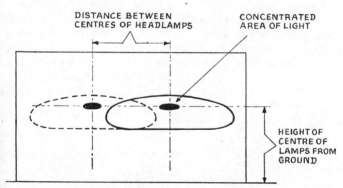

FIG. 171. Illustrating method of aligning headlamps (Austin)

Beam Alignment Instruments. The accurate alignment of headlamp beams can be carried out in the service station by means of special instruments or equipment, much more expediciously than the road method previously described. These instruments—known as Beam setter, Headlamp Aligner, or Tester etc.—consist of optical systems, mounted on portable stands or carriers, such that they can be wheeled up to the vehicle's head-lamps and adjusted to the correct vertical height, in each case. Provision is made for the correct alignment of the axis of the optical system with the longitudinal axis of the vehicle.

In the more recent Lucas Beam setter, the optical tube, of rectangular shape, is fitted at the front end with a condenser lens and at the other end with an adjustable projection screen. The condenser lens, when aligned with the headlamp front lens, projects an image, of reduced size, of the beam, thus giving a high intensity image on the screen. At the centre of the screen, a small hole is located and this allows light from the image to pass through and on to a photo-electric cell, which becomes energized allowing a current to flow through a special type of meter which is calibrated in candle-power (candelas) units. This gives a measure of the light intensity of the headlamp beam. The head lamp is then adjusted in the horizontal and vertical senses to give the highest candle-power reading on the meter.

The optical arrangement of the Lucas circular-type Beam setter is shown in Fig. 172. The screen on the right hand side is provided with black crossed lines, which can be observed through an aperture immediately below the candle-power meter. The equipment includes a special optical alignment bar having feeler arms for making contact with the rim of the rear wheel of the vehicle. The tube, having a reflector, light bulb and lens, throws an image

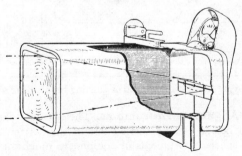

FIG. 172. Optical system of Lucas headlamp beam aligner unit

THE LIGHTING SYSTEM 277

FIG. 173. Method of using headlamp beam aligner.

on to a screen, having cross lines, located near the front wheel and the relative alignment of the front and rear wheel is adjusted until the image is at the centre of the crossed lines on the front screen. The axis of the optical tube is then parallel to that of the vehicle. The beam-setting optical unit, on the same stand as the wheel alignment screen can then be moved vertically or laterally to its testing position over the headlamp lens, so that its axis is always parallel to that of the vehicle (Fig. 173). It is possible for a trained operator to test both headlamps of a car in 5 to 8 minutes.

American Headlamps. A special alignment and beam-setting device, known as an *aimer* is used, in pairs, for setting the light beams. Each aimer is mounted direct on to the front lens of the headlamp, being held by suction cups and located by the three ground pads on the front of each lens. The slope of the floor on which the car stands is measured with the apparatus shown in Fig. 174 (A), and a suitable adjustment is made on each aimer. The aimers are provided with horizontal, or 'Left-Right' and vertical, or 'Down-Up' scales. When aligned for a test the Left-Right scale being then at zero, the line of sight is as indicated in Fig. 174 (B) and a split image of the target is seen. The horizontal adjusting screw on the headlamp is then adjusted until the split-

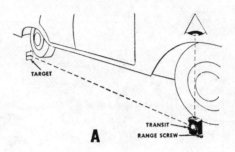

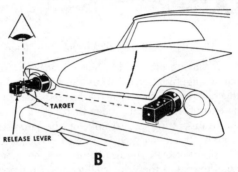

Fig. 174. Beam alignment of American sealed-beam headlamps (Chrysler). (A) determining the slope of the floor. (B) mounting and adjusting the beam aimers

image merges into one unbroken line. For the vertical adjustment a spirit level on the aimer is used; the vertical adjusting screw on the headlamp is turned until the bubble is central in its scale; then the headlamp is aligned vertically.

Dual Headlamp Units. The dual-type headlamp consists of two standard unit headlamps mounted side by side in a single housing (Fig. 175) enclosed by a single headlamp cover; a dual unit is fitted on both sides of the car. The inner or inboard lamp unit is fitted with a single filament lamp bulb of 50 or 60 watts, while the outer or outboard unit has the usual two-filament bulb,

THE LIGHTING SYSTEM 279

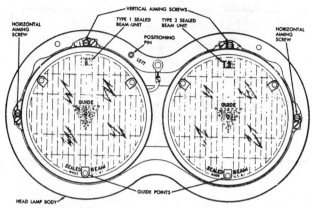

FIG. 175. Dual headlamps, showing vertical and horizontal alignment screws and the three locating or guide points on fronts of sealed-beam units

namely, one for the main driving beam and the other for the downward non-dazzle beam. When the foot-operated pedal switch is depressed, the downward beams of the outer lamps are brought into action, while the horizontal or driving beams are put out of action. For normal driving purposes the upper or driving beams of all four headlamps are in operation.

The lamp illustrated in Fig. 175 is of the double T-3 sealed-beam unit kind, each unit having three precision-faced projections or guide points, for beam alignment purposes, when a beam aligner or aimer apparatus is used in the servicing garage. Provision is made for aligning or aiming the driving beams on the dual headlamp, by means of two vertical aiming screws; a horizontal adjusting screw is provided for sideways alignment of each lamp. It will be seen that this is a similar arrangement to that of the single headlamp as shown in Fig. 163.

The headlamps can be aligned or aimed by projecting the individual beams from each lamp bulb filament on to a vertical wall or screen, at 25 feet ahead of the lamps. The car should be on level ground, with its tyres inflated to their correct pressures and the normal passenger and fuel load in the car. The screen projected light patterns for the outer units in the lower beam position are shown in Fig. 176 (A), and for the inner lamp, on the

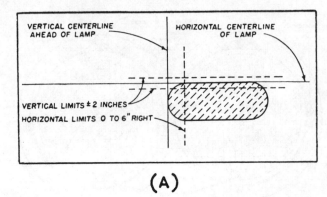

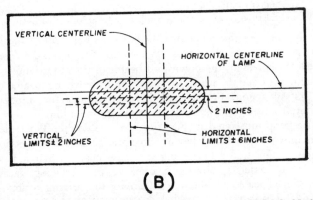

FIG. 176. Method of aligning dual headlamp beams (A.M.C. Nash)

upper or driving beam position, in Fig. 176 (B). The vertical centre lines shown in the diagrams should correspond with the vertical plane through the mid point between the inner rims of each dual headlamp and the distance between the right and left hand centre lines should be the distance between the mid points of each dual headlamp.

In reference to Fig. 176, it should be emphasized that the projected light patterns refer to American driving practice in which the driver is on the left hand side of the car.

THE LIGHTING SYSTEM

Fog Lamps. Experience of fog driving conditions indicates that the ordinary main driving beam is practically ineffective in penetrating the fog, the general result being a backward reflection of light from the fog particles. The dipped beam, however, proves more effective, especially when it illuminates the nearside of the road including the edge or kerb.

The better practice, with two filament headlights is to switch off the main beam filaments and drive on the dipped beam, in fog. Since, however, there is no adjustment provided for increased nearside direction of the dipped beam, it is generally considered better to use a special fog lamp which can be adjusted to give any nearside illumination required. The flat-topped beam, with reasonable side spread and dipped to the best inclination to give sufficient forward illumination while showing the road edges is provided in the well-designed fog lamp. However, while no really satisfactory lamp has yet been developed for denser fog conditions the more promising fog lamps appear to be the high candle-power, single filament type having a shield over the upper half of the bulb—to cut off upwardly reflected rays—and means for adjusting the lamp beam both vertically and horizontally. A satisfactory commercial fog lamp which employs this principle is provided with one vertical and two horizontal adjusting screws for directing the dipped beam. The lamp is adjusted to give the brightest intensity beam just below the flat cut-off line of illumination.

Another foglamp employed the principle of splitting the reflector in halves (Fig. 177) and then moving the lower half slightly forward. The lamp filament was located between the two focal points of the reflectors. This arrangement results in the light in the horizontal plane being spread sideways and the remainder reflected downwards, thus giving a wide beam having a sharp upper cut-off together with satisfactory downwards distribution. Further, this type of lamp was provided with a spherical shield at the upper back part of the bulb, as shown in Fig. 177, to cut off extraneous light.

In regard to the *mounting of fog lamps*, the official regulations in this country lay down that the fog lamp must be mounted on the vehicle not less than 2 ft. above the level ground surface when

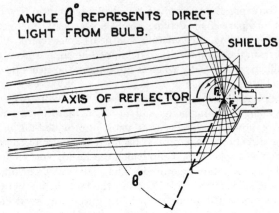

Fig. 177. Divided reflector fog lamp optical system (E. A. Watson)

the car is loaded and also that an observer whose eye level is $3\frac{1}{2}$ ft. from the ground must not be dazzled by the lamp, when he is at 25 ft. from the lamp; this condition also applies to dipped headlamps and pass-lamps.

The pass-lamp resembles the fog lamp and is sometimes made as a combined fog and pass-lamp; it has been used for illuminating the road when passing other vehicles but, with the introduction of the modern high voltage two-filament headlamp that gives the diverted or dipped beam when meeting other vehicles, the pass-lamp is not now favoured.

The long-range driving lamp is used, for fast night driving, as a supplement to the main headlamp beams. The lamp uses a special conical bulb-shield and crystal clear lens which produce a pencil type of beam of great penetration. This type of lamp is used with the two headlamps and not usually alone.

The Side Lamps. The side and tail lamps are miniature head-lamps with reflectors and opalescent or fluted covers to disperse the light. In some cases the side lamp bulb is mounted through a part of the front lamp reflector—a method enabling economy in production cost to be achieved. The side and tail lamps are not only required under driving conditions, but when a vehicle is parked at night. The lamps should not be less than 1 foot from the extreme sides of the vehicle nor less than 15 ins. or more than 6 ft. above the ground. The minimum illuminated areas of these lamps is that of a circle of 1 in. diameter. The bulbs used for side and tail lamps have 6 watt filaments.

Reflectors. A more recent regulation requires the fitting of a pair of red light reflectors at the rear end of the car. The minimum reflecting area allowed is that of a circle of $1\frac{1}{2}$ in. diameter. The reflecting properties must be such that if placed 100 ft. away from a white light source throwing a beam of light of intensity, 2,000 candelas in the direction of the reflector, the latter when turned in any direction at not more than $22\frac{1}{2}$ degrees must reflect a beam of red light of not less than 1/1000 candelas in any direction up to $3°$ from the light source-reflector direction.

Brake Warning Light. When the foot brake of a car is applied a switch is closed to complete the circuit of two rear warning red lights, indicating to rear traffic that the car is slowing down. Usually, a two-filament bulb is used in each of the tail lamps. The brake filament is 21 watt and the tail filament 6 watt. Provision is made in the bulb holder for preventing the bulb from being inserted incorrectly.

In American practice the brake warning light filament has a rating of 32 watts and the other parking filament, 4 watts. It is usual, also, to have a *parking brake indicator light* on the instrument panel.

Instrument and Indicator Lights. The illumination of the instrument panel is, usually, by small screw-cap bulbs of 2·2 watts rating. The lights used for indicating the oil pressure (low), battery discharge (dynamo not connected *via* cut-out), headlamp upper beam and the turn or direction indicators are usually of 2·2 watts rating, but British flasher light indicator bulbs are usually of 0·75 watts. American cars usually employ bulbs rated at 2 candle-power for indicator lights on the instrument panel and for such other purposes as the radio dial lamp, glove box lamp, etc. The ignition lock lamp is generally of 1 candle-power.

The Ignition Warning Light. This red indicator light on the instrument panel, as mentioned earlier, shows if the battery is discharging, when the ignition has been switched on. This discharge occurs when the engine is at rest or idling at a speed below that of the cut-out contacts opening. The warning lamp is connected in parallel with the series winding of the cut-out contacts unit. As the series winding of the cut-out is connected to the 'live' terminal of the dynamo brush the warning lamp is connected between this terminal and the ignition switch terminal, as shown in Fig. 178. When the cut-out contacts are open, battery current flows through the switch (closed), the warning lamp and dynamo to earth. When the dynamo speed is such that the cut-out contacts close, the red lamp is virtually out of circuit as the dynamo current then flows through the cut-out series winding through the switch and ammeter to the battery.

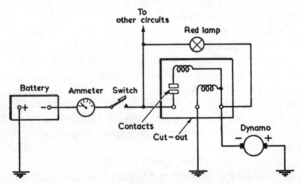

Fig. 178. Circuit arrangement for the ignition warning light

Reversing Lamps. These lamps are fitted as standard on certain cars and as optional equipment on other cars. They are brought into action when the lighting system is in use, by a switch which is actuated by the reverse gear lever movement and are now legal, in this country, despite the previous regulations concerning the display of only red lamps at the rear of the car. Only one reversing lamp may be fitted and the total power of its bulb, or bulbs, must not exceed 24 watts. Further, the lamp should not dazzle a person standing at 25 ft. from the car whose eye-level is not less than 3 ft. 6 in. above the level ground. Usually, the reversing lamp has a fluted-type lens or cover, to disperse the rear beam laterally so that the driver can see the full road width.

Automobile Lamp Bulbs. The earliest electric lamps on cars used carbon filament bulbs, but owing to vibration and voltage variation effects, these were too fragile. They were replaced by the metal filament (tungsten) bulbs which gave much better results. These vacuum-filled bulbs gave about 1·5 watts per candle-power for a filament temperature of 2,300° C but, later, were replaced by the gas-filled metal filament bulb, using the inert gas, argon, which used rather less than 1 watt per candle-power. The coiled filament bulb has proved satisfactory in modern automobile lamps for accurate alignment and low current consumption; it operates at a filament temperature of 2,600°–2,800° C and uses about 0·7 watt per candle-power.

Direction Indicators. Over a comparatively long period, in this country, the semaphore type of direction indicator was a standard fitment. It was mounted 4 to 5 ft. above the ground and on the sides of the car, so that an illuminated red or amber arm was shown on the side, to which the driver intended to turn. The hinged arm carrying its lamp was actuated by a small solenoid, when the driver's switch was closed. Usually, when the steering wheel had been brought to its central position after making the indicated turn, the current supply to the indicator arm was switched off automatically, by means of a switch on the upper part of the steering column, thus giving a self-cancelling operation to whichever arm was operated. A pilot light on the instrument panel showed that the direction indicator was working.

Following earlier American practice, the *flashing-type of*

286   MODERN ELECTRICAL EQUIPMENT

direction indicator is now standard on British cars, since the bright flashing lights at the front and rear of a vehicle can be more readily discerned than the relatively small constant light semaphore indicator.

The flashing lamps, however, must be wired so as to operate independently of the lighting system, i.e. in daylight as well as at night. It was the usual practice to fit a separate, but higher candle-power filament in a double-filament bulb fitted to the front side or parking lamp, namely, a 21-watt filament, whereas only a 6-watt filament is used for the side lamp. The rear flashing light is obtained from a separate 21-watt bulb in the rear red lamp, or more recently is fitted to its own amber-coloured rear lamp, so that this lamp will flash independently of the tail or brake-stop lamps.

There is also a driver's indicator lamp which flashes in unison with the front and rear flashing lamps and, in some cases, a series of 'clicks' is heard by the driver at the flashing frequency. In a later type of flashing lamp equipment the driver's signal operating lever, located just below the steering wheel, has a small light bulb at its outer end, which fluctuates in intensity when a right or left turn is being indicated.

The principle of the electrical unit which produces the inter-

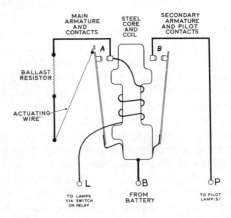

FIG. 179. Principle of Lucas flashing lamp direction indicator

THE LIGHTING SYSTEM 287

mittent current supply to the flasher bulb filaments is shown in Fig. 179 and the external complete circuit is shown in Fig. 180. The flasher unit has two sets of contacts A and B, an electromagnet, ballast resistance and a control or actuating wire. The

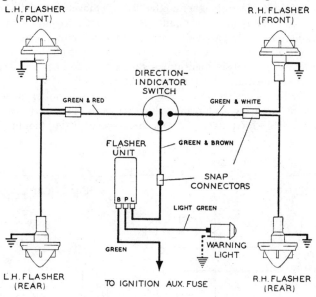

Fig. 180. Circuit diagram for flashing lamp system

operation of the unit depends upon the linear expansion of this control wire when heated by the current that flows through it. When the direction indicator lever is moved to the right or left current flows from the lower battery-connected terminal B through the control and ballast wires, the flasher lamp filaments to earth.

The ballast resistor limits the indicator current, so that the lamps do not light when the contacts A are open. However, the heating effect of the current expands the actuating wire which causes the contacts to close and thus allow the lamps to light. At the same time the control wire is short-circuited, causing it to cool down. It therefore causes the contacts (A) to open, when a current pulse again goes to the flasher lamp and the repeated cycle produces a series of flashes which in practice are between 60 and 120 per min.—as legally stipulated. Usually, however, the frequency is 70–80 per min. After the contacts (*A*) have closed, the current passing to the flasher filaments flows through a small coil in the base of the unit and the resulting electromagnetic effect causes movement of the iron core; this closes the upper contacts (*B*) (Fig. 179) and completes the circuit to the driver's switch indicator lamp, which then illuminates. When the control wire current is cut off, by the cooling action previously mentioned, the current to both the flasher filaments and the indicator lamp is also cut off.

The Lucas flasher unit is a small cylindrical member enclosed in a metal casing and provided either with three terminals or three plugs—for plugging into a three-socket fitting mounted on

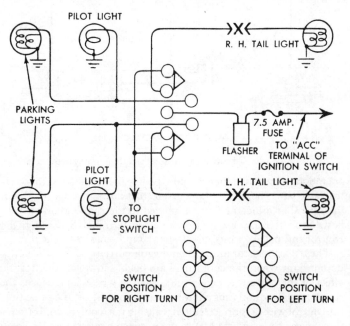

Fig. 181. Combined parking light and flasher filament bulb electrical system (Ford, U.S.A.)

the car. This method enables the unit to be removed for inspection or replacement purposes. The flasher indicator bulb is usually a 12-volt, 2·2-watt screw-cap (M.E.S.) bulb.

In some lighting systems the flasher filaments are located in the same electric bulbs as used for the side, or parking light filaments. In this connection a typical electrical diagram for the parking, front and rear lamps and flashing filaments is given in Fig. 181. This diagram shows also the direction indicator switch and its left and right turn positions and also the flasher unit.

Electronic Headlamp Control. In order to save the driver the trouble of depressing the foot switch for lowering the headlamp beam, for approaching traffic, this operation can be performed automatically, with the aid of a photo-electric device which is actuated by the light received from the oncoming cars.

The General Motors 'Autronic-Eye' headlamp control, which is fitted to Cadillac, Buick and other cars, automatically switches the upper beam filament off and the lower beam filament on, during night driving. The complete assembly consists of four separate units, namely, a photo-electric tube which picks up the light beams from the approaching cars, an amplifier, power relay and combination dimmer-override foot switch (Fig. 182).

The photo-electric tube (phototube) is energized by the forward illumination received from oncoming vehicles and the current is amplified and supplied to the power relay which is a heavy-duty relay which switches the headlamp filaments from the upper driving to the lower non-dazzle beam, whenever the light received by the photo-electric tube is sufficient; weaker illumination will not actuate the relay. When the oncoming vehicles pass the headlamps, the photo-electric tube current diminishes and the relay switches the lower beam filament 'off' and the upper beam filament 'on'.

The override section of the pedal switch operates only when the switch is in the automatic position; its purpose is to provide the upper beam by overriding the automatic control. This arrangement also permits the driver to signal with his headlamps, and enables the automatic system to be tested.

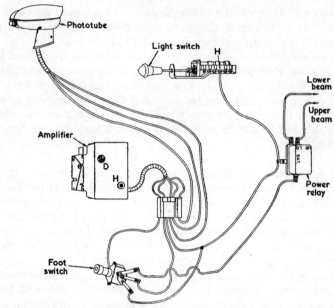

FIG. 182. The Buick electronic-controlled headlamp system showing the various components and connections

The equipment is provided with a 'driver-sensitivity' control. This device, which is located at the rear of the photo-electric tube, enables the sensitivity to be adjusted to compensate for the different amount of light coming through a clear or tinted windscreen. It also enables the relay switch to operate, for lowering the beam when an approaching car is *nearer* to the driver's car or *farther off*.

Fig. 183 illustrates the later but similar Buick Guide-Matic automatic headlamp control circuit diagram. It includes the photo-electric (phototube), driver's control for sensitivity, the amplifier and power relay electrical units and the driver's foot-pedal switch.

The automatic control headlamp equipment is adjusted on the car by means of special alignment or aiming apparatus, specially constructed for this purpose.

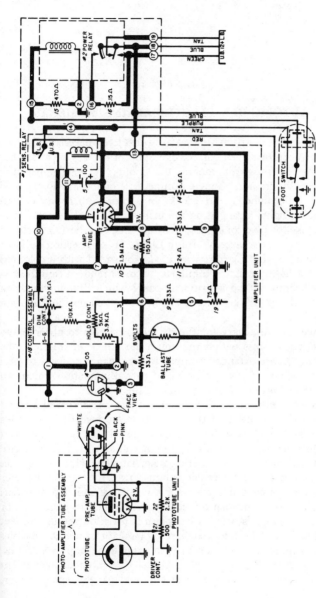

FIG. 183. Typical circuit diagram for electronic headlamp control (Buick Guide-Matic)

CHAPTER 10

AUTOMOBILE ELECTRICAL INSTRUMENTS

THE essential instruments that are considered necessary to keep the driver informed of correct operation of the engine and other systems or components include the Cooling Water Temperature Gauge, Oil Pressure Gauge, Ammeter, Main Tank Fuel-Level Gauge, Speedometer and Clock. In addition, there are certain indicating lights arranged on the instrument panel, which sometimes take the place of separate instruments, e.g. the Oil Pressure and Current Discharge Warning Lights. The Direction or Turn Indicator, of the Flasher type, is also provided with a coloured indicator light, or lights. Similarly, on most modern cars there is a small lamp on the instrument panel which illuminates when the Headlamp driving beams are in the raised position, i.e. not dipped.

Since it is necessary for the driver to be able to observe any of the instruments with a minimum movement of the head the instruments are arranged compactly on a panel in front of him and, for night driving, are provided with illumination which can be switched on or off by a switch on the facia panel.

More recently, in British practice, there has occurred a trend towards the mounting of the essential instrument dials or scales behind separate sectors on a large oval or circular mount, instead of employing separate dials for the instruments, which comprise the thermometer, oil-pressure gauge, ammeter and fuel gauge. The outer part of the circular mount is used for the speedometer scale, and the central portion for the mileage and trip recorder.

Since in some mass-production cars, neither an ammeter nor a pressure gauge is fitted, their place is taken on the instrument panel or mount by coloured warning lamps; this method enables a more simple combined instrument panel to be employed. Thus, as shown in Fig. 184, the AC-Delco combined instrument includes the water temperature and fuel gauges, speedometer,

AUTOMOBILE ELECTRICAL INSTRUMENTS 293

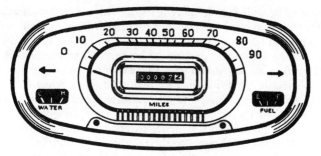

FIG. 184. The AC-Delco instrument panel (Vauxhall)

mileage recorder and indicator lights for the direction indicator, oil pressure and battery current discharge.

It is not usual on mass-production cars in the less expensive class to provide graduated scales for the thermometer, ammeter or pressure gauge—when fitted. Instead the needle positions are denoted by letters, e.g. in the case of the fuel gauge by E (empty) and F (full) and for the thermometer C (cold), N (normal) and H (hot).

Fig. 185 shows the American Lincoln car instrument panel,

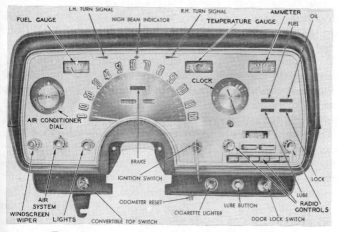

FIG. 185. An American instrument panel (Lincoln)

which is provided with more gauges, controls and indicating lights than in the previous example. The additional items include the air conditioner controls, for refrigerated and also heated air ventilation of the car; parking brake application warning light and warning or signal windows of rectangular shape—as shown on the middle right side, in Fig. 185—to indicate when the fuel is low in the main tank, low oil pressure, power lubricator not operating and the electric door locks, when in operation. The lubrication and door-lock switches are shown below on the right. The 'odometer reset' refers to the resetting knob for the mileage recorder of the speedometer.

Trends in Instrument Design. The general trend in automobile instrument design—as is shown by the descriptions that follow—is towards the *replacement* of the *mechanical* or mechanically-connected types of instrument by the *electrically-actuated ones*. Examples of instruments that can be replaced by electrical types include the radiator thermometer of the vapour pressure kind; the fuel gauge of the liquid type; the Bourdon pressure gauge and, more recently, the engine revolution and speedometer instruments. In this way more or less delicate metal tubes, rotating shafts and the use of volatile liquids are avoided, their places being taken by relatively small cables that can be located more conveniently. Again, metal tubes and the speedometer flexible cable drive casing are apt to become noisy.

When considering the replacement of mechanical by electrical instruments, the questions of ease by mass production and of accuracy of indication should be taken into account, since it is possible to replace any of the mechanical types by electrical equivalents, but not always more economically.

The more important instruments here considered include the gauges for measuring the fuel level in the main fuel tank, cooling water temperature, oil pressure and battery charge or discharge current. Instruments for measuring engine and road speeds, and for showing the mileages covered are also dealt with.

Printed Circuits. A development in the use of printed circuits which were used, originally, for radio receivers is that of a printed circuit for completing the individual circuits of all the lights and instruments in the instrument panel assembly of an automobile. In the later Buick cars, this printed circuit has circuit connector pins and a keyway to ensure the correct assembly of the disconnecting plug for the connecting pins. In the event of faults developing a new printed circuit panel can be used. This circuit includes all the instruments shown on the instrument panel and the various indicator lights.

Fuel Gauges. The earlier type of fuel gauge indicated the fuel level in the main tank by the height of a red-coloured liquid in a vertical glass tube on the instrument panel. The gauge operated upon the hydraulic head principle and necessitated a fuel pipe connection to the tank unit from the gauge.

The liquid-type gauge was, later, replaced by the electric fuel gauge, consisting of a *tank* or *sender* unit, connected by an electric cable to the instrument panel *indicator* unit, which was of the moving iron ammeter pattern.

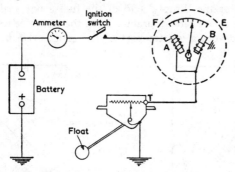

Fig. 186. Illustrating principle of electric fuel gauge

The principle of this fuel gauge is indicated in Fig. 186, in which the tank unit T is shown below and the indicator unit above, on the right hand side.

The tank unit consists of a float mounted at one end of a hinged arm which carries, at its upper end, a contact or slider unit which moves along a resistance member. When the tank is empty, the slider moves to the left, thereby inserting all of the resistance into the electrical circuit. When full, the resistance is practically zero. The current from the battery flows *via* the ammeter and ignition switch to one coil A of a moving coil ammeter, from which it divides, to flow through the other coil B and tank unit to earth.

When the tank is empty, the resistance of coil A, together with the full resistance of the tank unit, will be much greater than the resistance of coil B, so that the greater current flowing through B will attract the moving armature of the ammeter needle unit and move the needle to the right, i.e. to the 'E' (or empty) position. When the tank is full the slider cuts out practically all of the tank unit resistance so that all the current flows through the coil A, direct to the tank unit 'earth'; the needle therefore moves over to the left or 'F' (or full) side.

Usually, the float arm spindle centre is offset from the resistance unit's centre, in order to give a variable rate of resistance change, for angular deflection of the float arm. In this way it is possible to obtain suitable gauge indicator divisions, and also wider divisions near the 'Empty' end of the fuel gauge scale; in this connection it appears desirable, from the motorist's viewpoint, always to devote a greater part of the gauge scale towards the

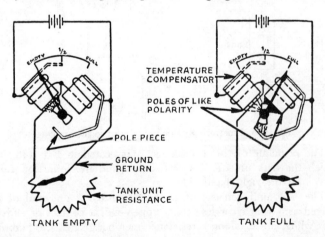

FIG. 187. Electric fuel gauge of typical American car (Oldsmobile)

AUTOMOBILE ELECTRICAL INSTRUMENTS 297

'Empty' end than that of the 'Full' end; usually, about one-third of the scale would be used to denote one-quarter of the tank's fuel contents.

Fig. 187 shows, schematically, the electrical circuits of a typical American fuel gauge for both the empty and full positions. The float arm resistance slider type of tank unit is employed in conjunction with a moving iron ammeter, having a pair of coils at 90° inclination, with an armature-pointer unit mounted at the inter-section of the coil axes. Silicone fluid is used in the armature bearing to damp out road vibration effects. One end of the left coil is connected to one pole of the battery while the other end connects with the end of the right coil and also one end of the variable resistance of the tank unit. The other end of the right coil connects with the other battery pole and the variable resistance slider; both connections are earthed or 'grounded'. The operation of the gauge is the same as for the fuel gauge previously described, but in this example a bimetal-type temperature compensator is shown; this compensates for atmospheric temperature changes. Fig. 188 shows the tank resistance unit and slider of this type of fuel gauge.

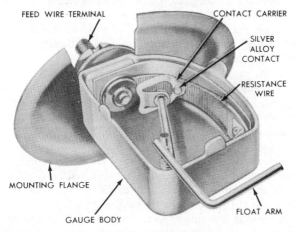

FIG. 188. Tank resistance unit and sliding contact, operated by float arm (Chrysler)

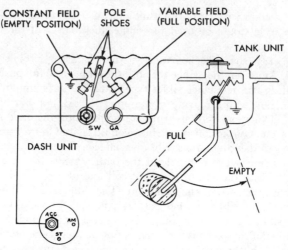

Fig. 189. Electric fuel gauge using constant current for one coil

In another version of the variable resistance tank unit type gauge, using a moving iron ammeter unit (Fig. 189), one of the instrument coils is energized with a constant current all the time, while the other coil is in series with the variable resistance of the tank unit, so that the magnetic field of this coil varies with the current in the variable resistance circuit and therefore controls the deflections of the gauge needle.

The principal disadvantage of the moving float fuel gauge is that the gauge needle is liable to oscillate when the fuel surges in the tank. Sometimes this trouble can to some extent be overcome by special baffles in the fuel tank but it is difficult to cure in the case of the modern shallow type of tank.

The Bimetal Fuel Gauge. This alternative to the moving iron type indicator uses the same variable resistance tank unit as in the preceding example but employs a bimetallic strip, heated by the current through the tank unit circuit to actuate the gauge needle. It has the important advantage over the moving iron type indicator of providing an effective damping movement to the gauge needle, so that the readings are unaffected by fuel surges in the fuel tank. In a well-designed indicator unit 90 to 95 per cent of the full deflection is obtained in about 30 seconds.

It is necessary to supply a constant voltage to the bimetal indicator and fuel tank unit circuit, since the gauge readings are sensitive to voltage changes. In this connection the constant voltage regulator, described later, is standard equipment.

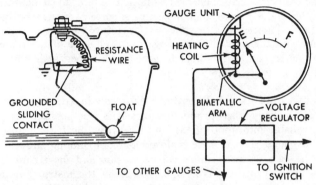

FIG. 190. The bimetal fuel gauge

Fig. 190 illustrates the bimetal fuel gauge circuit of a typical American instrument. It consists of the tank or sender unit on the left and the indicator or gauge unit on the right. The voltage regulator, below, is supplied with battery current *via* the ignition switch, the constant voltage current being also tapped off for the other electrical gauges on the automobile.

As shown in Fig. 190, the float is near the bottom of the 'Empty' tank and the rheostat slider has inserted all of the resistance into the gauge unit's circuit, so that only a small current can flow through the heating coil of the bimetallic strip and there will therefore be no deflection of the strip; the gauge needle is therefore arranged to read the E or empty scale graduation. When fuel is poured into the tank the float rises and moves the sliding contact upwards, cutting out more and more of the tank unit resistance. This results in a continuous increase in the current flowing through the heating coil, so that the bimetallic strip heats up and deflects towards the right at its lower free end.

This movement causes the gauge needle to move to the right, i.e. towards the 'F' or full reading side. Since the bimetallic arm alters its temperature rather slowly, the needle does not show any sudden movements due to fuel surges.

It is usual to provide a temperature compensator, in the form of a bimetal element, so that atmospheric temperature changes do not affect the gauge needle readings. The bimetal element then maintains the needle anchorage in its correct relative position and uninfluenced by atmospheric temperature changes.

As with the moving iron gauge unit, the scale of the bimetal fuel gauge can be modified to give a more open scale towards the 'E' or empty end.

Low Fuel Level Warning. With the inexpensive gauges fitted to many mass-production cars it is not possible to give any reliable indication of the fuel level near the E (empty) end of the gauge, so that there is always a risk of running out of fuel. In the case of many more expensive cars and most American cars an indicator light or signal panel on the instrument unit shows when the fuel level has become low; this is illustrated in Fig. 185. The indicator lamp is in series with the battery and a pair of 'low-level' contacts in the fuel tank.

Fuel Tank Safety Measure. In later designs of fuel gauges it is the practice to earth the free end of the fuel tank gauge resistance, to avoid any risk of sparks in the petrol-laden air in the tank.

The Constant Voltage Unit. As mentioned earlier it is necessary to maintain the voltage of the bimetallic type of fuel gauge constant to ensure accurate readings; the same requirement applies to the bimetal engine cooling water temperature gauge. In the case of 12-volt systems it is usual to regulate the instrument circuit voltage at 10 volts. For the previous 6-volt systems the regulated voltage was 5 volts. The constant voltage regulator consists of a bimetallic arm surrounded by a heating coil, as shown in Fig. 191. When the battery current flows through the coil the bimetallic arm becomes heated and bends, thereby causing a pair of contacts at the free end of the arm to

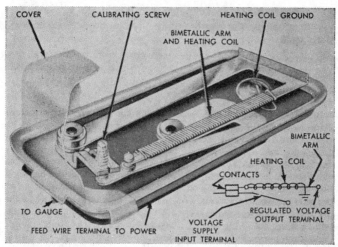

FIG. 191. The constant voltage regulator, for instrument supply purposes

open and thus break the coil circuit. The current supply ceases and the arm cools down to bring the contacts together again, when the arm becomes heated and the cycle of circuit break-and-make is repeated. This succession of circuit changes causes a pulsating voltage which provides an average of 10 (or 5) volts to the instrument circuit. The constant voltage unit (Fig. 191) is provided with a calibrating screw which enables the rate at which the circuit is made and broken to be changed, in order to give the required 10 (or 5) volt average. Usually, this setting, after being made by the manufacturer, is sealed, to prevent interference.

The Engine Temperature Gauge. Previously, the temperature of the engine cooling water was measured in the cylinder-head part of the water jacket by means of a temperature unit consisting of a small steel bulb, containing a volatile liquid, connected to an instrument panel located gauge by means of small bore steel tubing. When the jacket water temperature increased the vapour pressure created in the bulb was communicated to the gauge—which was of the Bourdon tube pressure type but calibrated in temperatures. Thus, by measuring the vapour pressure, the corresponding temperature was indicated on the dial.

While this kind of gauge worked satisfactorily and gave accurate readings it was often difficult to install and liable to damage of the capillary tubing and so was subsequently replaced by the electrical types of thermometer of which there are two principal alternative types, namely, (1) *The Thermal* and (2) *The Semi-conductor* systems.

(1) The Thermal Gauge. This earlier temperature gauge operated on the principle of a stationary contact and a moving contact mounted on the free end of a bimetallic arm, as in the voltage regulator previously described. The cylinder head unit contained the bimetallic arm with its heating coil and contacts. When the cooling water was cold the contacts vibrated at a certain frequency but as the temperature was increased the tension between the contacts was gradually reduced and the contact closure period became shorter, causing a higher vibration frequency. The temperature gauge was of the bimetallic arm kind, heated by the same current as the cylinder unit. As the water temperature increased, the greater the frequency of vibration of the contacts resulted in current (or energy) being applied to the gauge bimetallic arm for a relatively longer period, so that the arm deflected to a greater extent and thus moved the needle over towards the 'Hot' side of the scale. Owing to the thermal inertia, or lag, of the gauge bimetallic arm the effects of the rapid vibration of the contacts were unable to cause needle vibrations, so that a steady reading was obtained. No voltage regulator was required with the thermometer. This type of thermometer had certain practical drawbacks associated with the contact surfaces; moreover it was more complicated and costly than the types which superseded it.

(2) The Semi-Conductor Thermometer. This instrument has a cylinder head, or sending unit containing a semi-conducting element in the form of a pellet made of a special material possessing the property of increasing the electrical resistance as the temperature is lowered and reducing it, with increase of temperature.

In a typical example, a pellet having a resistance of 140 ohms at 15° C, would have a resistance of 93 ohms at 50° C; 70 ohms at 100° C and 60 ohms at 150° C. This reduction in resistance with temperature increase is made use of to indicate engine cooling water temperatures by the increase in current in a constant voltage circuit.

There are two methods by which the indicator unit or gauge can make use of this semi-conductor element property, namely, (A) by using the circuit current to operate a moving iron type of ammeter or, (B) using a bimetallic arm to actuate the gauge needle.

(A). *The Moving Iron Indicator*. The circuit diagram of the cooling water thermometer is shown in Fig. 192 in which the

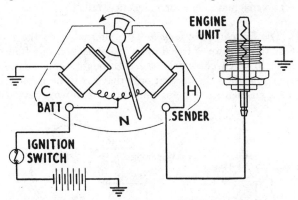

FIG. 192. The engine cooling water temperature gauge, for semi-conductor engine plug-unit and moving iron ammeter (Vauxhall)

cylinder head or engine unit, of the semi-conductor kind, is indicated on the right and the gauge unit, with its pair of inclined coils on the left. Since this unit is practically identical in principle to that shown, previously, in Fig. 189, it is unnecessary to describe it again. The operating current is supplied from the battery *via* the ignition switch so that the thermometer *only*

304 MODERN ELECTRICAL EQUIPMENT

operates when the *ignition is switched on*. In this application the current through the left coil is constant throughout the application of the gauge, while the current through the right coil changes according to the resistance of the pellet in the engine unit. When the cooling water is cold the current from the battery will flow through the left coil, to earth and the pole piece of the left coil will then attract the pivoted armature, bringing the needle N to the 'C' or cold side of the temperature scale.

When the engine unit pellet begins to heat up, its resistance will diminish, so that the current through the right coil will increase. The magnetic field will therefore become stronger, so that the armature will move in the direction shown by the arrow and the needle will move from the 'C', or cold position, towards the 'H', or hot end of the scale.

It should be explained that both coils have a similar polarity at their upper faces, and the armature responds to the resultant of the two magnetic fields or magnetic 'pulls'.

(B). *The Bimetal Arm Gauge*. This alternative to the moving iron type of indicator is based upon the deflection of a bimetallic arm or strip when current is passed through a coil wound over the arm. Since it is the current only that has to be measured, or indicated, an identical type of gauge to that shown previously in

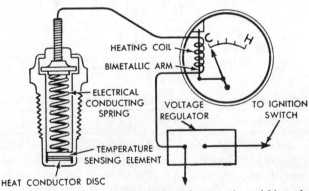

FIG. 193. Temperature gauge for semi-conductor unit, and bimetal arm indicating gauge (Ford)

Fig. 190 can be employed; the scale will, of course, be calibrated to read temperatures in the present case.

Fig. 193 shows, schematically, the temperature gauge circuit for a typical bimetallic arm gauge unit, in which the heating coil circuit contains the temperature sensing element, i.e. the semiconductor pellet with its conducting spring, the voltage regulator and (not shown) the ignition switch and battery; it will be seen that the gauge section is identical to that of Fig. 190, but with a different bimetal unit and scale.

Water Temperature Warning Lights. In the case of Buick and similar cars, a pair of signal lights on the instrument panel indicate the permissible water temperature limits. A cylinder head unit comprises a temperature switch having double contacts. When the water temperature is below 110° F a pair of contacts close to light a *green signal*. The light goes out at 110° F and remains out until a temperature of 245° F is attained, when the other pair of contacts close and illuminate a red window on the instrument panel. The temperature 245° relates to that of the pressurized cooling system.

Oil Pressure Indication. There are two methods of indicating whether the oil pressure in the engine lubrication system is satisfactory, namely, (1) The Instrument Panel Warning Light and, (2) The Oil Pressure Gauge. The former method is used on some mass-production cars as it is less expensive than the pressure gauge and takes up less room on the instrument panel. The latter method gives a continuous reading of the oil pressures and is to be preferred.

(1) The Oil Pressure Warning Light. In this method, when the ignition is switched on and the engine is at rest or idling, a coloured light—usually green—is shown on the instrument panel; this indicates that the oil pressure is low, or there may even be no pressure at all. If, however, when the engine is accelerated to at least 1,000 r.p.m., the light goes out, this shows that the oil pressure has increased above the lower limit of the coloured light switch. This switch is opened by a pressure-sensing device, such as a diaphragm or piston, when the oil pressure reaches a given value. In most instances this limiting value is 8 to 12 lb. per sq. in. The pressure unit is connected to the 'delivery circuit' of the lubrication system at the most convenient place on the crankcase unit and a single cable connects the insulated terminal to a small lamp holder—usually of the 2·2 watt type and thence to a 'live' battery connection. Thus, the pressure unit switch, indicator lamp and battery are wired in series (Fig. 194).

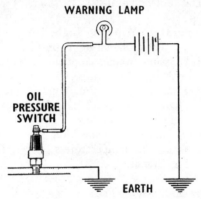

FIG. 194. Oil pressure system warning lamp circuit

The operating pressure can readily be checked by placing a Bourdon-type pressure gauge in the circuit and noting its reading when the light goes out.

(2) The Oil Pressure Gauge. This is invariably of the non-electrical Bourdon gauge type and consists of an elliptical-section tube bent to the arc of a circle. The tube is connected to the oil pressure system. It is a well-known property of this curved elliptic tube that when subjected to internal pressure it tends to straighten itself, but as one end of the tube is fixed, the other end only, moves outwards. It is thus possible by means of a simple linkage to connect the free end of the tube to the gauge needle and in this way measure the pressure on a scale which is a linear or equal division one, since the movement of the end of the tube is proportional to the pressure within the tube.

AUTOMOBILE ELECTRICAL INSTRUMENTS

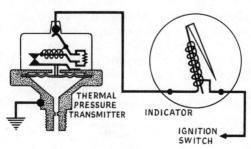

FIG. 195. Oil pressure gauge, using vibrating contacts and bimetal gauge unit (Smith)

The diameter of British pressure gauges is 2 in. and the scale occupies 55° of arc. These gauges are accurate to within about 3 per cent of the true pressure values.

The chief drawback of the Bourdon gauge as used for automobile engines is that of the connecting tube, between the engine and instrument panel, which is liable to leakage at the joints —since the rubber-mounted engine movements require flexibility of the tube—and to fracture in servicing the engine, when *in situ*. It may also give rise to noises.

For these reasons an electrical type of gauge would appear preferable. In this connection gauges under development operate upon the sender or transmitter principle of vibrating contacts (Fig. 195) or the variable resistance method used in fuel gauges, but using a pressure-sensing device to vary the resistance and a bimetal indicator unit identical to that used for fuel level and temperature gauges, as described earlier.

The Ammeter. Automobile ammeters are usually of the centre-zero pattern in which the right side of the scale shows the *charging current* and the left side, the *discharge current* of the battery. As mentioned earlier, the starting motor is not included in the ammeter circuit, since the starting currents are of the order of hundreds of amperes, whereas the usual ammeter range is minus 30—zero—plus 30 amperes.

The ammeter used is of the inexpensive moving iron type having a pivoted iron armature and a permanent magnet control.

It is provided with a special damping device since, otherwise, the readings would be affected by the vibrations of the voltage and current regulator contacts.

Fig. 196 shows the arrangement of a moving-iron ammeter fitted with both inertia and friction damping. For the former result the armature is coupled to a properly balanced weight which can rotate through a limited arc. The coupling consists of

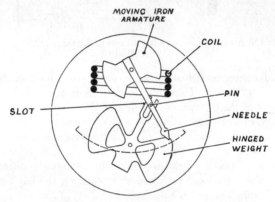

FIG. 196. Principle of the moving iron ammeter

a pin on the ammeter needle which engages with a slot in an arm of the balanced weight; this method provides a damping effect which is proportional to the vibrations or movements of the pointer.

The ammeter is desirable as an indication of the battery charging and discharging conditions and in connection with any adjustments that may be required to the voltage or current-and-voltage regulator to alter the regulated current values. In many mass-production cars, however, the expense of an ammeter is avoided by arranging for a red lamp indicator to be illuminated on the instrument panel whenever the battery is discharging, e.g. when the engine is at rest or idling slowly with the cut-out contacts closed. A typical red lamp indicator circuit is shown in Fig. 178. The lamp used is similar to that of the oil pressure warning light, namely 2·2 watts with screw M.E.S. cap.

AUTOMOBILE ELECTRICAL INSTRUMENTS 309

The Speedometer. While the greater part of the speedometer system is purely mechanical the actual needle-actuating system is electrical, but there is an alternative to the usual mechanical speedometer, namely, the all-electrical system referred to later.

The conventional speedometer consists of a flexible shaft drive taken from the rear end of the gearbox, off the output shaft, through a worm gear unit, to the speedometer unit on the instrument panel. Thus, the speed of the drive shaft is proportional to the road speed of the car, assuming no tyre slip.

The car speedometer unit operates on the magnetic induction principle, in which a drag cup made from a low resistance metal, e.g. aluminium, is rotated by the drive shaft. Inside the cup is a permanent magnet attached to a separate shaft. As the cup rotates,

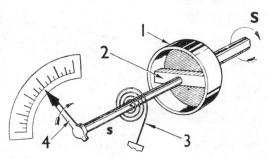

FIG. 197. Principle of magnetic induction speedometer. S—drive shaft. (1) drag cup. (2) permanent magnet. (3) control spring. (4) needle. s—instrument shaft

the field from the magnet induces eddy currents in its rim and these produce a torque or drag tending to pull the magnet round in the same direction as the cup is moving. This rotation is resisted by a control spring; the greater the speed of rotation, the greater is the torque, the scale being a linear one.

Referring to Fig. 197 this shows the drag cup (1) rotated by the shaft S — on the right — while the magnet (2) is attached to the instrument shaft s by the spring (3) and it carries the needle (4). It will be seen that as the shaft rotates in a clockwise direction it tends to drag the magnet around, thus giving a needle deflection on the speedometer scale.

The speedometer driven shaft also actuates a trip and total mileage recorder, on the well-known bicycle cyclometer principle. The trip operating mechanism is usually, but not always, in modern production cars, provided with a push-pull knob below which is used to reset all of the trip figures to zero on its scale.

Horizontal Scale Speedometer. These more recent speedometers which are fitted to American and certain British cars operate upon the same drag-cup principle but the movement of the drag cup is utilized to partly rotate a horizontal drum, or roller, behind a long horizontal slot in the instrument panel. The drum is provided with a datum line or a coloured sector which moves along the fixed scale on the upper edge of the slot, in order to indicate the road speed.

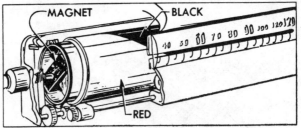

Fig. 198. The horizontal scale speedometer (Buick)

In the case of the roller-type speedometer (Fig. 198) the roller is arranged behind the horizontal slot carrying the fixed scale and a helical coloured part on the roller moves along the scale. The intersection of the farthest point on the oblique helical line and the scale gives the road speed reading.

Fig. 199 shows the AC-Delco speedometer and instrument panel of the more recent Velox cars. The lower illustration shows the total mileage recorder, and the speed drum. Up to 30 m.p.h. the colour of the scale is green but between 30 and 60 m.p.h. it changes to orange and above 60 m.p.h. to red. The reading shown on the upper dial is 60 m.p.h. The dial also shows the fuel and temperature gauge readings and the usual indicator or warning lights. The housing of the unit is an integral casting into which the various components are assembled.

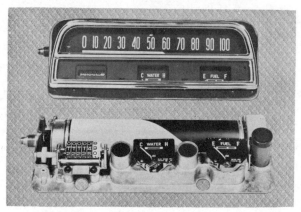

Fig. 199. Instrument panel and assembly, showing horizontal scale speedometer and mileage recorder, etc. (Vauxhall)

The Electrical Speedometer. The principle of this type is that of an electrical generator driven off the gearbox output shaft, i.e. at a speed which is a function of the road speed, which supplies current to a voltmeter, graduated in miles per hour; since the generator voltage is practically proportional to its speed the scale of the voltmeter is almost a linear one.

A typical arrangement for an automobile speedometer is one using an A.C. type generator designed to give the required voltage range over the rotor speed limits so that the sensitivity of the moving coil voltmeter used as the road speed indicator is not appreciable and the needle range over the road speed scale is adequate. Since the current generated is alternating it is necessary to provide a selenium- or silicon-type rectifier inside the casing in order to convert the A.C. to D.C. for actuating the moving coil voltmeter. In order to calibrate the instrument there is, in the case of the Smith electric speedometer a graded resistance in series with the moving coil. A typical Smith movement, in production, has a 25 milliampere full scale deflection giving an almost linear scale with a range of 250°. The generator provides 1 volt per 100 r.p.m.

The electrical layout of the Electromag electric speedometer,

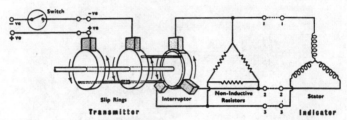

Fig. 200. Illustrating principle of the Electromag electric speedometer (Leyland)

which is used on Leyland vehicles is shown in Fig. 200. The complete assembly comprises a transmitter and indicator, or speedometer dial unit, together with the necessary connecting cables.

The *transmitter* unit which is driven by a flexible shaft from the gearbox, consists of a die-cast frame having a shaft on which are carried two slip rings and a commutator-type interrupter or rotary switch. Current from the vehicle's battery is supplied to brushes which bear on slip rings. The interrupter slip ring has three brushes which have three non-inductive resistors connected between them, as shown in Fig. 200.

The *indicator* is a self-starting synchronous motor consisting of a stator and rotor; the stator has three slots carrying field coils to form a three-phase winding. The rotor has two magnetic elements, one being rigidly fixed to the shaft and the other free to rotate on the shaft but held by friction to the other element. The rigid element is a permanent magnet which provides the synchronous torque; the other element has a high hysteresis value and contributes the starting torque.

An extended part of the motor shaft carries a worm which, through suitable gearing drives a permanent magnet which energizes the drag cup, as in the previously-described speedometer. The worm gearing also drives the mileage counter.

The drag cup and its pointer are mounted on a spindle, which is restrained by a spring, the torque due to the drag cup being directly proportional to the speed of rotation of the permanent magnet.

Speedometer Developments. It will be evident from these brief considerations that the present trend in speedometer design is to provide an alternative all-electric system that will be both compact and relatively no more expensive than the gearbox speedometer drive and indicator. The provision of mileage recorders, similar to those of mechanically-driven speedometer units, must however be taken into account when assessing new electrical designs.

The Engine Tachometer. This instrument which indicates the engine revolutions per minute is usually fitted to high performance cars, e.g. sports cars; it enables the driver to check the engine speeds under all conditions of gear changing and general driving. It operates on a similar principle to the shaft-driven speedometer, but special more accurate models are often fitted.

Ignition-type Tachometers. Use is made in this type of speedometer of the frequency of the ignition sparks, thereby obviating the use of a separately driven generator, as in the preceding example. The low amperage current pulses due to the high voltage sparks occur at a frequency proportional to the sparks and, therefore, to the actual engine speed. By means of a special electrical circuit these signals are transformed into a direct current supply which is indicated on a moving coil voltmeter, as before. The Cirscale revolution indicator, of the Record Electrical Company, London, works on this principle. The advantages of this method are the elimination of the usual gearbox-driven speedometer shaft with its flexible casing, in the case of 'mechanical' speedometers or the generator of the all-electric speedometer, so that the only units are the electric converter for the D.C. supply and the voltmeter.

Electric Clocks. The electric clock, operated by battery current has largely replaced the spring-driven type. It uses very little current and its accuracy of time-keeping is independent of normal voltage variations although very high or low voltage values will produce some effect on the accuracy.

These clocks may, however, show small variations over long

Fig. 201. The electric impulse clock (Smith)

operating periods but provision is made for altering the clock's hands to correct or re-set the time. There is also an adjusting device to alter the clock rate; this is at the back of the clock in some instances and at the front in other cases.

The electric impulse clock is not self-starting but is readily re-started by pushing in the re-set stem and releasing it again (Fig. 201).

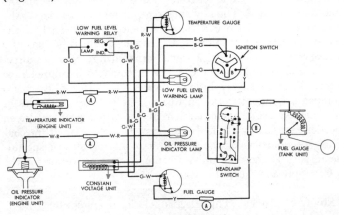

Fig. 202. Wiring diagram for the instruments and constant voltage unit of automobile

Electrical Circuit for Gauges. The general principle adopted when wiring up an electrical gauge is to connect the live terminal of the gauge to the most convenient live terminal of a nearby electrical accessory or junction unit and to earth the other gauge terminal. A typical example of the method of wiring the gauge is shown for the Lincoln and Continental car, in Fig. 202. This diagram shows also the internal wiring of the temperature, pressure and fuel level gauges as well as that of the constant voltage unit, referred to earlier in this chapter.

The low fuel level warning relay and warning lamps and also the wiring of the headlamp switch unit are also indicated. The letter pairs on the cables refer to the distinguishing colours used to identify these cables, e.g. W–R (white red band), B–G (black-green band) and so on.

The initial live current supply is taken from the ignition switch which has three positions, namely, 'Off', 'Ignition On' and 'Starting Motor On'.

CHAPTER 11

MISCELLANEOUS ELECTRICAL EQUIPMENT

APART from the electrical components previously dealt with there are several other items of electrical equipment on modern automobiles that are fitted as either standard or optional. In this connection the tendency on British mass production cars, other than those in the lower price class, is to fit as standard equipment the following: The Horn or Horns, Windscreen Wiper, Windscreen Washer, and Car Interior Heater and Demister. In British Motor Corporation cars the electric fuel pump is used instead of the mechanical engine-driven type.

The electrical equipment of standard American cars is more extensive, being apparently based upon the principle of providing everything possible, within certain limits, for the driver's and passengers' comfort—and ease of control, in the case of the driver. Reference to some of these fitments is made later in this chapter.

It is becoming the tendency in more recent cars of most nationalities to employ electrically-operated devices in place of mechanically-controlled or similar alternatives, so that an increasing burden is placed upon the battery and its charging systems; as stated earlier the charging dynamo or generator of an automobile has to provide a much higher current output than hitherto.

It is proposed to describe briefly some of the important electrical accessories that have not already been dealt with in previous chapters, commencing with those commonly employed on British cars.

Windscreen Wipers. The earlier hand-operated single windscreen wipers were later replaced by the electric-motor and vacuum types employing a single or pair of interconnected wiper blades to provide a field of clear vision for both the driver and passengers. Of the various kinds of wipers that have been used,

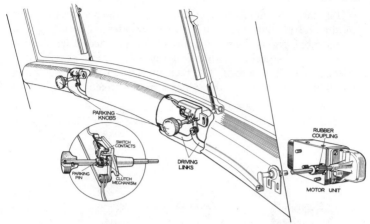

FIG. 203. Arrangement for two independent windscreen wipers

the dual type shown in Fig. 203 was popular for a time. It consisted of a pair of wiper blades or arms pivoted at their lower ends in fixed bearings housed in mountings behind the facia. Each blade was provided with a short lever, these being connected together by a connecting rod. The right hand lever was also connected to a parallel lever, on its right, which was given a rocking movement by means of a shaft extending forward into a gearbox unit containing an electric motor of the shunt-wound pattern having a single field coil and a reduction gear to slow down the output shaft of the gearbox to about 30 to 40 r.p.m. The output shaft was provided with a crank at its end, forming the previously mentioned right hand lever. A rubber coupling at the gearbox end prevented noise transmission to the driven mechanism. Each wiper arm unit was provided with its own clutch, and the right hand clutch also embodied the switch for starting the wiper motor. The action of pushing in and turning the knob placed the driver's blade on the screen, engaged the dog clutch, and switched on the motor. The passenger's blade was engaged by its own dog-clutch, but it could be disconnected and parked on one side, leaving the driver's wiper blade in operation. When both blades were parked the hinged windscreen of that period could be opened.

The later development of the windscreen wiper uses a wiper motor and gearbox (Fig. 204) which actuates a special kind of metal cable, giving the cable a to-and-fro, or linear reciprocating movement. The cable is wound tightly with a helix of stiff steel wire so as to form a rack. The rack engages with pinion gears in auxiliary gearboxes and the wiper arms are fixed at their lower

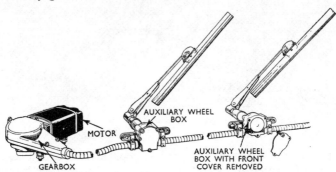

FIG. 204. Flexible cable rack windscreen wiper, showing actuating gearbox and motor unit on the left hand side

ends to the shafts of the gears (Fig. 204). Thus, as the metal cable-rack reciprocates it gives each of the pinion gears a rocking motion, which is transmitted to the attached wiper arms; the rate of wiping action is about 40 oscillations per min.

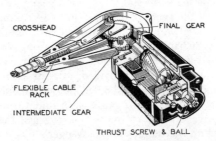

FIG. 205. Motor and gearbox, showing reduction gear mechanism

Referring to Fig. 205 the motor shaft is provided with a worm gear meshing with the smaller worm wheel. This gear has a

MISCELLANEOUS ELECTRICAL EQUIPMENT 319

smaller pinion attached to it, which meshes with a larger gear above, so that a double reduction ratio effect is obtained. The upper face of the larger gear is provided with a crankpin which is connected to another pin on a slider block attached to the end of the steel cable by a short connecting rod. The slider block is thus given a reciprocating movement which is imparted to the cable so that the latter moves to and fro lengthwise. It should be emphasized that the cable rack does not rotate, although it is flexible and enclosed within a strong but flexible outer casing; in the improved model the cable rack operates in rigid tubes located between the motor and the two wiper-arm gearboxes. Further, there is a special *circuit breaker* of the current-operated thermostatic kind which protects the motor against serious *overloading*, as when the blades encounter ice or hard snow on the windscreen. When the overload is removed the motor resumes its normal operation.

Parking of Blades. When the current is switched off, at the instrument board switch, the wiper blades return, automatically to the lowest right or left hand side of the windscreen where they leave the windscreen free from obstruction. When switched to the parking position the direction of the motor's armature is

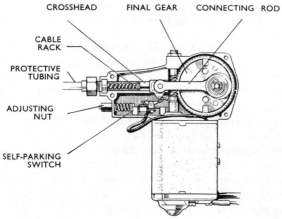

FIG. 206. The adjusting nut for the self-parking switch (Humber)

reversed, this movement causing an eccentric between the crankpin and connecting rod to increase the movement of the slider block or crosshead. The latter unit carries a striking arm which then opens the self-parking switch (Fig. 206) which cuts out the motor. The self-parking switch is adjusted for correct operation by means of the adjusting nut shown in Fig. 206.

Central Position Motors. Instead of using a motor driving unit located at one end of the windscreen, a single unit can be mounted under and behind the centre of the windscreen. The motor then drives through a reduction worm gear to a pair of larger gears provided with crank pins which reciprocate two short levers to and fro but in opposite directions. These levers are connected by connecting rods to similar levers which work the wiper arms. This arrangement enables much shorter connecting rods or links to be used, and generally results in a more compact unit.

Variable Speed Models. More recent car windscreen wipers are provided with a two-speed motor, so that when the wiper blades have to deal only with rain on the screen the higher wiping speed is used, but when snow or icing conditions are encountered the lower speed is employed, since a greater effort is then required. Similarly, when the wipers are used on a dry or drying screen more frictional resistance is encountered and the lower speed is then needed.

The Lucas DR3 two-speed windscreen wiper uses the same mechanism to operate the wiper arms as that shown in Fig. 206. The operating motor, however, is provided with a three-position rotary switch marked 'H' (High), 'N' (Normal) and 'P' (Park), giving the high speed, low speed and parked positions.

The electrical diagram for the two-speed motor and the switch positions is reproduced in Fig. 207; in these diagrams the moving contacts within the switch are shown by the dotted lines. The negative current from the 'live' battery terminal is supplied to the motor armature by way of a thermostatic circuit breaker; the field current is also supplied *via* the thermostatic circuit breaker to terminal (8) on the switch. The positive (earth) side is connected to terminal (13).

MISCELLANEOUS ELECTRICAL EQUIPMENT 321

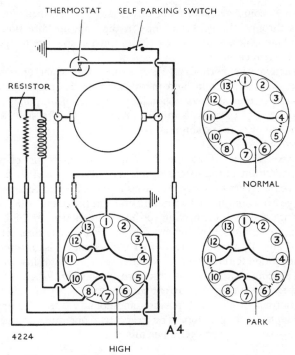

FIG. 207. The internal connections of the two speed wiper motor and switch (Humber)

The higher motor speed is obtained by inserting a resistor in series with the field. Thus, with the 'H' position of the switch, current goes to the field circuit by way of the terminal (8) which is then connected internally to (7) and (10).

The lower or normal motor speed corresponding to the 'N' position of the switch is given by the current being supplied to the field *via* terminal (8) which is connected internally to (10). The terminals (4) and (5) are interconnected to (1) and the resistor is now out of the circuit, so that the voltage is applied across the field winding only, thus giving the lower motor speed. In the parking or 'P' position the switch disconnects the armature from its earth terminal (1), the current then being supplied *via* (8) which is connected internally to (5) and (6). The current now flows through the field windings in the opposite direction thus enabling the gearbox parking switch shown in Fig. 207 to operate, as explained earlier.

When in the normal or 'N' position, the wiper operates at 45–50 strokes per min. and in the high or 'H' position, 60–70 strokes per min. The corresponding current consumptions of the motor are 2·7–3·4 and 2·6 amperes, respectively. The stall current of the motor is 10–11 amperes.

Three-Blade, Two-Speed Wiper. A later model wiper designed to meet the special problem of cleaning a wide, shallow and acutely curved windscreen uses a powerful two-speed motor and gearbox unit having a rotating crank at the output side, which is connected by a connecting rod to a hinged arm on the central windscreen wiper arm mount. The hinged arm is thus reciprocated and, by means of two other connecting rods transfers the same motion to the two outer wiper arms. The motor has a similar self-parking switch to that described previously. When operated on a 12-volt supply the current consumption is 2·5 to 4 amps. for the normal speed switch position and about 2·0 to 3·5 amps. for the high speed position. The respective operating frequencies are 45–50 and 60–70 cycles per minute.

Windscreen Washers. In the later model of these previously hand-operated washers an electric motor is employed to pump water from a container through a flexible tube to a pair of jets on to the windscreen. Known as the Screenjet (Fig. 208) the self-contained unit is enclosed in a glass container in which the washing liquid is stored. It consists of a small permanent magnet type electric motor which is mounted inside and at the top of a moulded cover that screws on to the top of the glass container. The armature shaft drives a vertical shaft at the lower end of which is a small centrifugal pump enclosed in a small auxiliary reservoir. The reservoir pump and its drive shaft are mounted inside a vertical brass tube which is secured at its upper end to the inside of the moulded cover. When the motor is switched on the water which enters from the container to the auxiliary

MISCELLANEOUS ELECTRICAL EQUIPMENT 323

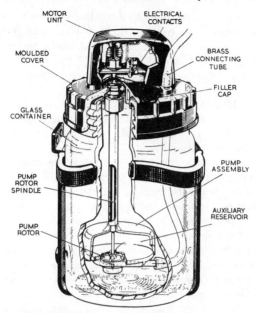

Fig. 208. Windscreen washer unit with internal motor driven pump and automatic sprayer device (Lucas)

reservoir passes through a fine gauze filter into the base of the pump, to the delivery side of which is attached a flexible tube which is connected to a brass tube fixed through one side of the moulded cover. From the outside end of the brass tube another flexible tube conveys the water under pressure to a pair of calibrated nozzles mounted on the metal bonnet support, just below the windscreen.

The washer is fitted with an automatic control, such that on depressing the dashboard switch the water is sprayed on to the windscreen for a limited period, namely 7 seconds, which is followed by an interval and then another spraying period. This continues until the reservoir is emptied. The upward thrust due to the centrifugal pump rotor is employed to move the rotating parts upwards, this movement being used to close a pair of contacts inside the motor housing, and to shut off the water inlet holes in the base of the pump. The closed contacts enable the current transmitted to maintain the motor in operation until the contents of the reservoir are discharged, when the water inlet holes are again uncovered.

The motor current supply is $1\frac{1}{2}$ amps. at 12 volts.

Electric Horns. The electrically operated horn, as a means of warning to road users, has entirely superseded the earlier hand-actuated devices, e.g. the rubber bulb, reed and trumpet and Klaxon diaphragm types.

In order to meet modern driving requirements a horn must satisfy certain acoustical and practical conditions. Thus the emitted note should be neither too musical, nor too raucous, i.e. such as to be objectionable to the public. Again it should operate over long periods without maintenance or adjustment attention. While it is possible to produce a horn that will penetrate to relatively long distances, such a horn would be objectionable to town dwellers. On the other hand a less penetrating but more mellow note horn would not be satisfactory for the purpose.

It has been shown that a horn emitting *low frequency notes* of between 250 and 500 vibrations per second, while absorbing appreciable electrical energy, has a very good distance penetration, but when used under city traffic conditions its notes can be lost in the low frequency traffic noises.

On the other hand, a *high frequency note* horn, of 1,500 to 2,500 vibrations per second, can be heard distinctly above traffic noise conditions, although it has only a limited penetration, due to the low energy input.

The desirable characteristics of these two types can be combined in a single horn, using two diaphragms, namely one to supply the lower and the other the higher frequency notes as indicated in Fig. 209. The horn casing contains an electromagnet, within which is an armature which is attached to a wavy-section low frequency, i.e. 300 vibrations per second, diaphragm which is clamped to the casing around its edges. The armature rod extends beyond this diaphragm and, near its outer end is secured a metal 'tone' disc which is designed or tuned to give a frequency of about 2,000 vibrations per second. When the horn switch button

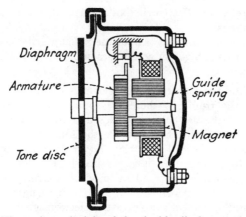

FIG. 209. Illustrating principle of the double diaphragm electric horn

is pressed the armature is set into vibration, by means of the contact unit (above the armature) which is struck by a projection on the armature, so as to break the circuit as the armature moves to the right; the current is then cut off and the guide spring shown moves the armature to the left so as to complete the horn circuit again. As the armature unit strikes the magnet system at the end of each stroke, overtone notes are produced by the tone disc; the frequency of these overtones is arranged to be an exact multiple of the low frequency diaphragm notes so that resonant high frequency vibrations are emitted by the tone disc. The resulting combination of high and low frequency vibrations produces an effective but not unpleasant note of good penetration.

The Double Horn System. Instead of using the single horn described, which may be regarded as a compromise between frequency requirements, it is more satisfactory to fit a pair of electrically energized horns having their frequencies tuned to a certain musical interval. In this connection tests have shown that the best musical interval is the *major third*. The horns are wired in parallel and are actuated together by a single switch. When the horns are provided with flares or trumpets, the air columns of which resonate at the same frequencies as the horns, the notes emitted are more effective in both penetration and combined sound qualities.

The horns are mounted behind the radiator grille and should not be obstructed by other accessories under the bonnet.

The Wind-tone Horn. The principle of this type of horn is that of the trumpet, where the sound is produced by the resonance of the air column enclosed in the trumpet. When applied to automobile horns, while the same trumpet or flare member is used, instead of setting the air column in motion by means of the human lips, it is initiated by the vibrations of a diaphragm which is energized by an electromagnetic unit. The actual note emitted depends upon the trumpet length and is not affected by the voltage supply to the electromagnet, or by any adjustment to the diaphragm. When the diaphragm is set into vibration the air column in the trumpet part is excited into resonance, so that it vibrates at the same frequency. It is thus possible to design an electric wind-tone horn to give not only the frequency note but also the proportion of harmonics in the air column so as to provide the desired quality of note.

As with the previous double horn method, the major frequencies of the twin wind-tone horns should differ by a major third.

Fig. 210 illustrates the Lucas Wind-tone horn, with its cover

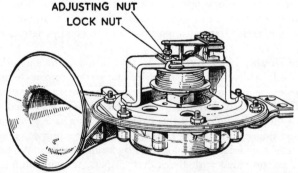

FIG. 210. The wind tone horn with its resonant trumpet (Lucas)

MISCELLANEOUS ELECTRICAL EQUIPMENT 327

removed to reveal the electromagnet and contact units; one of the contacts is adjustable by means of a screw-nut and lock-nut. As mentioned before, any adjustments that are made do not affect the frequency of the horn. The type of horn shown in Fig. 210 requires an operating current of about 6 to 7 amps.

Relay-type Horn Systems. In the case of certain types of electric horn, requiring relatively heavy currents, e.g. for 6-volt systems, the contacts of the horn switch would be apt to pit or erode, due to the sparking across the contacts as they operate. To obviate this, a relay device is fitted so that while the horn electrical system requires a larger current, namely, about 9 to 12 amps. at 12 volts, the horn button circuit current is relatively low, namely 3 to 5 amps.

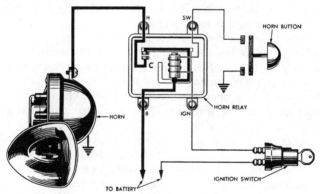

FIG. 211. The relay-operated electric horn (Dodge)

Fig. 211 shows the relay system used on certain Dodge cars, in conjunction with a wind-tone (Spartan or Auto-Lite) horn. When the horn button is pressed this completes a circuit containing the relay solenoid, ignition switch and battery. The solenoid is thus energized, causing its armature to move downwards bringing the contacts C to close to connect the battery directly to the live terminal of the horn. Since the other battery and horn terminals are earthed, the circuit is therefore completed and the horn will sound.

Notes on Electric Horns. In British practice the horn is wired independently of the ignition switch so that the horn can be sounded when the ignition is switched off. In many American systems the horn is wired through the ignition circuit and so can only be used when the ignition is switched on. (See Fig. 211.)

It is illegal in this country to sound the horn when the car or other vehicle is stationary. Further, it is an offence to sound a horn between 11.30 p.m. and 7 a.m. in a built-up area.

The Electric Fuel Pump. As an alternative to the engine camshaft-operated mechanical pump for delivering fuel from the main tank to the carburettor, the electrical version has advantages. Thus, it can be located away from the engine in any convenient position. It does away with the mechanical drive and engine-located pump parts. When it is necessary to inspect or maintain the electric pump it is more accessible for the purpose while, in the event of failure it can be removed quickly and replaced by another pump. It has the further advantage of operating independently of the engine, e.g. when the engine is at rest or during the starting or pre-starting period. The mechanical pump will only commence to operate when the engine is cranked, so that if there has been a fuel leakage in the system starting of the engine can be delayed appreciably.

With the mechanical pump it is usually necessary to prime the pump, after the engine has been standing for some appreciable time. This is effected by a small hand-priming lever on the pump. In this connection, the electric pump is self-priming.

The principle of the most widely used type of fuel pump, namely, that using a flexible diaphragm pumping element is illustrated in Fig. 212. The pump comprises the diaphragm which is clamped tightly around its edges and is flexed up and down by the plunger of a fixed solenoid which is energized from the battery when the ignition is switched on and the lower pair of contacts closed by the springs above the solenoid. The plunger is then drawn into the solenoid and the diaphragm is flexed downwards, thus creating a suction effect which causes the inlet valve to open inwards so that fuel, from the main fuel tank, flows into the enlarged shaded space, shown in Fig. 212. At the same time

MISCELLANEOUS ELECTRICAL EQUIPMENT 329

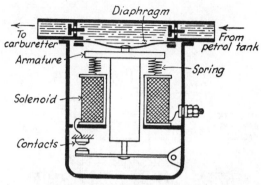

FIG. 212. Illustrating principle of electric fuel pump

the outwardly opening delivery valve is closed by atmospheric pressure in the carburettor float chamber. When the plunger moves downwards by a predetermined amount it forces the spring arm downwards and opens the contacts, thus cutting off the current to the solenoid. The springs then move the plunger upwards, to complete the delivery stroke, by flexing the diaphragm upwards, and the contacts are again closed, so that the pumping cycle is completed and then repeated as long as necessary.

This type of fuel pump has a delivery output of 8 to 13 gallons per hour, according to its size and a suction lift of 36 to 48 in., with a delivery height of 3 to 4 feet.

In modern cars the S.U. electric fuel pump,* which is fitted to all B.M.C. cars, is generally located at the rear of the car close to the fuel tank. It is necessary to provide sound absorption, otherwise the pump operation noise is clearly discernible.

The pump is fully satisfactory and reliable over long periods of service but it is necessary to ensure that the contact breaker points are cleaned and, if required, adjusted after appreciable service.

Another type of electric fuel pump that has been used employs a flexible bellows unit instead of a diaphragm, but the diaphragm type has now superseded it.

The more recent electric fuel pump made by Lucas Ltd.,

* Fully described in Volume 2 Motor Manual Series.

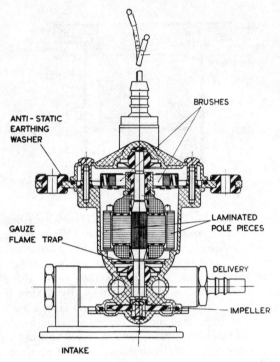

FIG. 213. Centrifugal type motor-driven fuel pump

illustrated in Fig. 213, employs an enclosed electric motor coupled to a small centrifugal pump. The complete and liquid-tight unit is submerged below the surface of the fuel in the main tank. Its main advantage lies in the fact that it is impossible for any vapour-lock to occur, since no partial vacuum can be created in the inlet side of the pump.

The motor is of the high energy permanent magnet field system kind supplied with its current by two P.V.C. covered supply cables which pass through petrol-proof plastic tubing and a special Hycar rubber gland, thus giving a completely petrol tight joint where the cables pass out of the unit.

Special precautions have been taken to avoid ignition of the fuel or its vapour by the brushes of the motor. The pump re-priming, should the tank become exhausted of fuel, is expedited by the provision of a small bleed hole in the pump casing, through which, during normal running a small discharge of fuel occurs.

When operating on a 12-volt supply, the pump will supply 25 gallons of fuel per hour with a relief valve-regulated pressure of 3·5 lb. per sq. in. for a power consumption of 18·3 watts, i.e. a current of about 1·5 amps. The normal speed of the armature is about 2,900 r.p.m.

Radiator Cooling Fan. The widely-used radiator water cooling fan which is driven by a vee-pulley (on the crankshaft forward end) and belt, has certain disadvantages so that more recently alternative methods have been introduced to overcome these drawbacks.

The cooling fan is necessary only when the engine is operating with the car at rest or moving at slower speeds, e.g. up to about 15 or 20 m.p.h. Above these speeds there is a sufficient flow of air through the radiator matrix to maintain the cooling water temperature at its correct value. The cooling fan is then not required.

With the belt-driven fan, the power required to operate it increases with the engine speed so that at the higher speeds several horse power may be absorbed. To obviate this disadvantage, the cooling fan can be driven by an electric motor which is switched on and off by a thermostatically-controlled switch. Thus should the water temperature increase above its normal value the thermostat located in the upper part of the cooling system acts through a relay device to switch on the fan motor automatically. The fan then cools the water to its normal value or range after which the motor is switched off automatically. In this way the cooling water is maintained at its correct temperature for maximum engine efficiency, and the power that otherwise would be absorbed by the belt-driven type of fan is saved.

The intermittent use of the electrically-operated fan results in greater fuel economy, less wear and tear, some increase in engine output at the higher speeds, a marked reduction in noise and quicker engine warm-up from cold.

332 MODERN ELECTRICAL EQUIPMENT

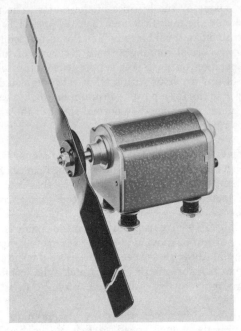

FIG. 214. The Lucas Type 3GM motor-driven radiator fan

The Lucas 3GM type electric motor fan (Fig. 214) which is mounted on four rubber units and protected by a splash-proof cowling has a current consumption, during its operating period, of between 6 and 7 amperes and weighs 4 lb.

Torque-sensitive Fan Clutch. Although not electrical in operation, mention should be made of a water-cooling system fan which is equipped with a temperature-sensitive coil which controls the flow of silicone fluid through a clutch, so that when the fan discharge temperature is low the fan clutch limits the speed to about 1,000 r.p.m. When the discharge temperature is high the temperature-sensitive coil actuates a port to allow a greater flow of silicone, so that the clutch is engaged fully to give the maximum air discharge, corresponding to a speed of 2,000 to 2,200 r.p.m.

MISCELLANEOUS ELECTRICAL EQUIPMENT 333

This type of fan clutch is fitted to more recent General Motors car engines.

Car Interior Heaters. While the car heater which is becoming the standard equipment on home and export cars is largely mechanical in its conception, it employs an electric motor driven centrifugal-type blower to circulate the heated air and also the unheated atmospheric air, under summer conditions. The heater unit consists of a heat-exchanger or radiator-type unit through the fine cores or matrix spaces of which heated water, taken from near the top of the engine cooling water system is circulated. Air from the atmosphere is drawn through the air passages in the radiator unit by either the forward motion of the car or by the electric motor-driven blower. This heated air is directed to ducts inside the car and to slotted ducts at the bottom of the windscreen —where it is used to demist or defrost the glass surface.

Heater fans for medium-size cars are driven by motors taking from 2·5 to 5·0 amps at 12 volts.

Car Ventilating Systems. For the complete air-conditioning of a car it is necessary not only to provide for interior heating of the interior under colder weather conditions, but also for interior cooling under hot weather conditions. For this purpose a refrigerating system is provided to maintain the interior air at 15° to 25° F below the outside air temperature. This air is circulated by a relatively powerful centrifugal fan driven by an electric motor through ducts in the rear part of the car. In a typical system suitable for the larger types of British and American cars an electric motor of 180 to 220 watts output delivers from 250 to 275 cu. ft. of air per minute into the car's interior. On a 12-volt supply the current demand of the motor, at full output, is about 15 to 18·3 amps.

It should be mentioned that the air compressor unit of the refrigerator system is driven by belt and pulley from the engine crankshaft pulley. Provision is made to disengage the drive by means of a mechanical or solenoid-operated clutch; this ensures that the compressor does not operate when the air-conditioning system is not in use.

Electrically-operated Windows. It has been the practice with most American cars to operate the front and rear door windows electrically, by means of suitably-located selection switches.

In addition some of the larger cars are provided with electrically-operated front ventilating glass side windows, or panels, and in certain instances the back window of the car.

Usually there are four main units employed in the electrical system, namely, a circuit breaker, a relay, an electric motor and a control switch.

There is a separate motor for each door window actuating mechanism and, since the control switch current would be too great for the usual type of switch a relay double-pole, double-throw switch is employed. This is energized by the hand switch and supplies current to the window lift motor.

Fig. 215 shows a typical window lift wiring diagram, for one door window. The window lift control switch is a low-capacity single-pole, double-throw type which is used only to actuate the relay. The relay contains a 30-ampere circuit breaker, which takes the place of a fuse, as explained in Chapter 12. The main circuit breaker is of the usual 50-ampere kind.

Referring to Fig. 215, when the door switch is moved in either direction the corresponding coil inside the relay is actuated and the electric motor in the door assembly is operated to raise or lower the window. It should be mentioned that the colours indicated in Fig. 215 refer to the cable colours. The main current supply to the relay is taken from the starter circuit solenoid. The motor employed is usually of the split-series type taking from 30 to 40 amps. and operating at 2,000 to 2,500 r.p.m.

There is a *master switch* on the left front door, consisting of a group of switches of the same type as those used on each door; each master switch unit is wired in parallel with the door switch it controls. Each door switch connects with the relay shown in Fig. 215, so that when the switch lever is moved up, the window is raised and when pushed down, the window is lowered. The window can be stopped in any desired position, by releasing the switch lever.

Usually, the door motor shaft has a worm fitted to it, which

MISCELLANEOUS ELECTRICAL EQUIPMENT 335

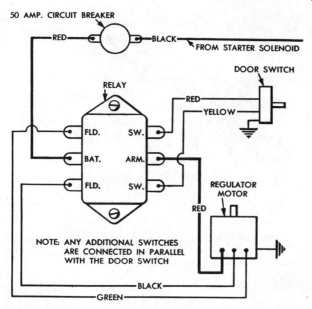

Fig. 215. Electrical circuit for four-window opening and closing mechanisms (De Soto)

engages with a worm sector, the gearbox for which is mounted on a plate unit. The partial rotation of the sector actuates the window lifting mechanism. A pre-loaded coil spring is fitted to the mechanism. In some instances the armature shaft worm engages with a smaller worm wheel which meshes with a large gear sector, so as to provide adequate leverage for window operation.

Fig. 216 shows the front door window lift mechanism of one model Lincoln car, the various components being indicated by the arrowed lines. The inclined motor is mounted on a common unit M with the worm-driven gear and larger gear sector. The rotation of this sector actuates the scissor-type mechanism shown. The opening and closing of the cross-arms lowers and raises the windows, there being two points of application to the window frame units; each comprises pivoted rollers. The rollers on each of

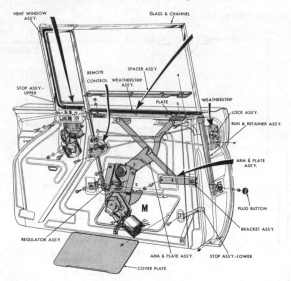

Fig. 216. Front door main and ventilation electrically-operated windows (Lincoln)

the two upper arms slide to-and-fro in roller guides built into the lower channel of the glass mount. The lower arm roller slides in the guide bracket, which is bolted to the inner panel of the door.

The same kind of mechanism is used for manually operated door windows. The front door assembly shown in Fig. 216 also includes a power-operated ventilation window, at the front end. The electric motor and gear assembly for this window is indicated on the left hand side, just below the window panel.

In some American cars the sector-shaped or 'quarter' window of convertible and other models, and the rear of the rear door window are arranged for power operation.

Power-operated Front Seats. Usually, power-operated seats are provided as an alternative to manually-operated ones, but in many American cars they are the standard equipment. According to the class or price of car, the power seat can be operated in four, five or six directions. Thus, in the four-direction type the seat can be moved backwards, forwards, up and down by means of a four-way switch control. In the six-way seat arrangement, in addition to the four movements just described the seat may be tilted up or down and the back rest tilted between two limiting angle positions.

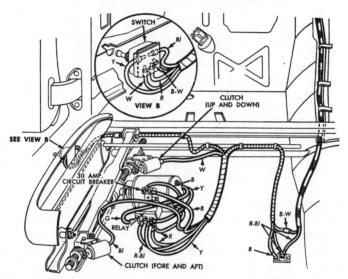

FIG. 217. Power-operated seat mechanism (Ford, U.S.A.)

Fig. 217 illustrates a typical front power seat arrangement, used in American Ford Company's cars. The power seat is controlled by a simple toggle switch having four selected positions. The seat mechanism is operated from a single electric motor which has a worm reduction gear which drives a fore-and-aft screw assembly (Fig. 218). There is a solenoid-operated clutch at each end of the screw, such that when the front clutch is engaged its nut-member is driven by the vertical end screw, to actuate a pair of pivot arms at each side of the seat to *raise* or *lower* the seat.

When the rear clutch is engaged, by means of the control switch, the rear or horizontal part of the driven screw moves the seat *forward* or *backward*. This fore-and-aft movement is trans-

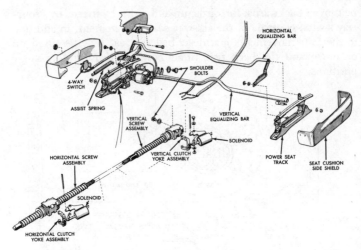

FIG. 218. Showing motor-driven seat mechanism in more detail

mitted to the right seat track by means of an horizontal equalizing bar. It should be mentioned that there is also a vertical movement equalizing bar.

Fig. 217 shows also the cable connections between the ignition switch, the 30 amp. motor circuit breaker, the operating switch, the electric motor and the two solenoids. The letters alongside the cables are for colour-identification purposes.

In the Chrysler power seat system which is of the six-position type previously referred to, an electric motor located beneath the seat, with its axis horizontal and arranged transversely, drives a gear train which operates transverse flexible cables that engage with and drive slave units on either side, below the front seat. These slave or power units actuate fore-and-aft gear racks and also subsidiary flexible drive cables to provide the various seat movements.

It is usually arranged with power seat systems that when the ignition is switched off, the front seat is moved to the rear limit position, so as to provide more space for the front seat occupants to leave, or enter the car.

MISCELLANEOUS ELECTRICAL EQUIPMENT

Convertible Body, Hood Operation. While in many European cars the convertible hood mechanism is operated manually, in certain instances, e.g. the English Ford cars the hood retracting and erecting mechanism can be actuated by an electric motor, which, through reduction gear and connecting linkage, is connected to the hood mechanism. The electrical circuit, which is a simple one, includes the 'up' and 'down' switch control, fuse (or circuit breaker) and the electric motor.

Retractable Hardtop. A more recent American Ford car is fitted with a combined mechanical and electrical system whereby the complete steel roof assembly, or hardtop unit can be raised bodily, moved rearwards and lowered into the rear luggage compartment. The lid (deck lid) of the luggage compartment must first be raised to allow the hardtop to enter and then closed again after the hardtop is stowed securely. The entire cycles of retracting and re-erecting the hardtop are performed automatically through electrically powered mechanical linkages.

The roof and deck lid mechanisms are driven by electric motors through flexible drive shafts which actuate screw jacks. Assist springs are attached to the linkage to help in the operating of the deck lid and to smooth out the final motion of the roof as it lowers in to the luggage compartment.

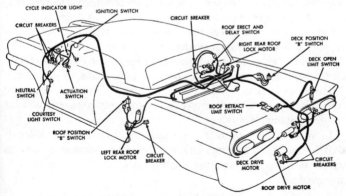

FIG. 219. Location of electric motors and switches for retractable hardtop operation (Ford, U.S.A.)

The electrical system is necessarily complicated as will be apparent from Fig. 219 which shows the locations of the electric motors, of which there are seven of the reversible type, and the initial actuation switch and some, but not all, of the eleven limit switches. There is also a neutral switch and a cycle-indicator light which is actuated by the unlocking of the deck lid. The electrical system is protected by a circuit breaker in the control circuit feed, a common motor feed circuit breaker and an individual circuit breaker in each motor earth (ground) feed.

When the car's transmission system is in neutral, power is then available for operation of the roof, through its actuation switch. This is a spring-loaded switch which returns to neutral when it is hand-released.

It may be mentioned that there are no less than fourteen separate electrical circuit diagrams for the various operations and the electric motors control and power circuits.

The current consumptions range from a minimum of 8 amps. to a maximum of 120 amps. according to the particular cycle or operation. The time of the complete retract or erect cycle is 45–55 seconds.

Electricity and Transmission Systems.[*] Electrical components, such as switches, solenoids, electromagnetic clutches, are used in certain semi- and fully-automatic transmission systems. Most overdrives for conventional type gearboxes also employ solenoids to engage the overdrive lower top gear ratio. There are, also, examples of gearboxes which use electromagnetic clutches in conjunction with epicyclic gear trains to obtain the different gear changes. In the Smith fully electrical automatic transmission use is made of excitation coils and magnetic powder for gear-changing clutches.

In most semi-automatic transmissions, gear lever or selector switches are used in conjunction with solenoid-actuated vacuum valves and also a carburettor throttle valve, to effect gear changes automatically.

[*] For fuller information, refer to *Modern Transmission Systems*, A. W. Judge, Motor Manual Series Vol. 5. (Chapman & Hall Ltd.)

MISCELLANEOUS ELECTRICAL EQUIPMENT

Miscellaneous Electrical Accessories. In modern automobiles electricity plays an important part in performing operations—merely by operation of switches—which were hitherto performed manually; in addition, new electrical items are appearing on new cars to increase their public appeal and to improve driving comfort and control. It is here possible to select only a few representative items of equipment from the many that are becoming available.

Electric Door Locks. On recent American cars the doors are locked by means of push-pull solenoids, attached by connecting rods to the door lock operating levers. It is necessary to operate a single-pole double-throw switch mounted on each of the front door trim panels to energize all four solenoids for locking, or unlocking the doors. Means are usually provided for locking or unlocking the doors both electrically and mechanically.

It is usual to fit a 'door-lock' indicator light on the instrument panel. In the case of the Lincoln and Continental car a single door lock switch is located on the intrument panel, to lock all four doors. This switch and its pilot light are shown in Fig. 185.

Driver Speed Warning and Control Devices. These devices warn the driver when a certain pre-set road speed has been attained and, in one instance, acts as an automatic speed control or regulator.

A typical arrangement, fitted to certain General Motors cars, e.g. the Buick employs a buzzer which can be set to operate at any speed between 20 and 110 m.p.h.; the adjustment is conveniently located for the driver, there being a knob on the left side of the instrument panel. The speed at which the buzzer is set is indicated by a dial on the left side of the speedometer dial.

The buzzer is wired in series with a 5-amp. fuse to the ignition switch on one side and the speedometer insulated ring at the back of the instrument panel, which has an edge-type dial graduated from 20 to 110 m.p.h. An earth (ground) brush projects from this insulated ring in such a position that an earthed plate on the *speedometer drag cylinder* will rotate with it as the cylinder rotates, thus earthing the circuit and actuating the buzzer. The insulated ring with its dial and earthing brush can be rotated by turning the re-set knob while observing the dial through the window on the left side of the speedometer dial.

The buzzer is a simple solenoid provided with an armature having a spring arm support; a contact at the end of the arm engages with another fixed contact, which is connected to earth through a 330 ohm, one watt resistor. The rapid make-and-break of the contacts provides the buzzer noise.

The Auto-Pilot device fitted to recent Chrysler Corporation cars is provided with a speed limiting device to indicate when a pre-set vehicle speed has been attained. It can also act as an *automatic speed control* to maintain the car at the pre-set speed.

The Auto-Pilot is governed by car speed and is driven by a flexible cable from the transmission speedometer drive pinion; this drives a spring-loaded governor which is controlled by a speed control knob on the instrument panel. A reversible electric motor is connected directly to the accelerator pedal linkage. When the pedal is pressed down, so that the reaction pressure is about 5 to 7 pounds, this reminds the driver that the pre-set speed has been reached.

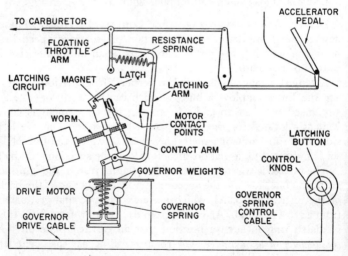

FIG. 220. The Chrysler Auto-Pilot road speed limiting device, shown schematically

MISCELLANEOUS ELECTRICAL EQUIPMENT

When the automatic speed control is required the car must be accelerated to the selected speed where the reaction pedal pressure is felt. Then, a 'lock-in' knob, located in the centre of the speed selector, is pulled out. The speed control device then comes into operation and maintains the car speed without pressure on the accelerator pedal. The automatic speed control is cut out of action by pressure on the brake pedal or by pushing in the 'lock-in' knob.

The regulating knob on the instrument panel operates a rack and pinion assembly which applies or releases pressure on the accelerator pedal at the pre-set speed. This pressure, when activated by the knob is applied to the governor. When the central 'lock-in' knob is pulled out and the car is moving the governor moves up and down on its shaft and positions itself when the car speed causes the governor forces to balance the spring forces as set by the speed control knob. The governor is linked to a contact arm which rotates to and fro between two contacts; these control the direction of the electric motor drive (Fig. 220).

The motor will rotate the worm shaft against a ball-nut and pintle assembly to open or close the throttle, according to the pre-set road speed to maintain the speed constant (Fig. 221).

In connection with the use of such speed indicating or automatic control devices, it should be understood that American highways and turnpikes provide very long stretches—often of hundreds of miles—so that it is possible to maintain fairly constant high driving speeds.

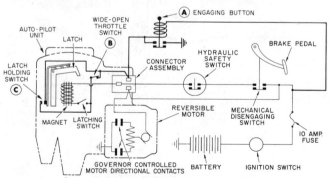

FIG. 221. The Auto-Pilot circuit diagram

Automatic Rear View Mirror Dimmer. The disturbing glare on rear view mirrors, at night, due to vehicle headlamps behind the car, can be reduced to a comfortable level by mirror tilting and similar manual-control devices. A more recent development utilizes a photo-electric cell and amplifier to tilt the mirror whenever a light beam of a given minimum intensity strikes the mirror. The cell is located at the back of the mirror and receives its activating light beam through a small aperture in the silvered surface of the mirror. The current is amplified and operates a solenoid unit to tilt the prism mirror slightly upwards; this provides a dim image to the driver. When the light intensity of the beam falls below a given value the mirror returns to its daylight driving position. The Mirror-matic device of Chrysler cars has an adjustable device so that the mirror can be made to tilt at whatever light intensity the driver requires.

Reserve Fuel Tank Control. When the fuel in the main tank falls below a certain level a reserve-supply can be brought into

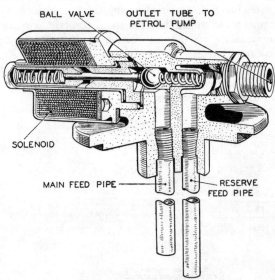

Fig. 222. Electrically-actuated reserve fuel tank supply (Lucas)

use by means of a solenoid controlled valve located at the top of the fuel tank, as shown in Fig. 222. In this case the tank is provided with two suction pipes of different lengths, and both pipes normally deliver petrol until the level of the fuel in the tank falls to the lower end of the shorter pipe. Air then enters this pipe so that the fuel flow ceases. When, however, this pipe is closed by operating a switch on the instrument board, the fuel will commence to flow to the fuel pump and carburettor. The effect of operating the switch is to cause a solenoid plunger to close a ball valve in its seating so as to stop ingress of air through the main feed suction pipe.

Radio Receivers. Fitted as optional equipment on some cars and standard on others these receivers take their electrical supply from the battery. A typical modern installation consists of a two-transistor push-pull audio output receiver, a transistor driver stage, a front and rear loud speaker with rear speaker fader control, a continuously variable tone control and a socket for connecting a car-type record player for a set of 10 to 14, 7-inch 45-r.p.m. records. In some models a foot switch is employed for 'search tuning', by depressing the switch it will select a station on the radio.

Certain cars are fitted with motorized aerials (antenna) at either the front or rear of the car, so that the telescopic aerial can be extended or retracted by a small electric motor, through reduction gears.

The power consumption of radio receivers fitted to British cars is from 35 to 45 watts.

Other Electrical Items. In the preceding pages some of the more important items of electrical equipment have been described, but necessarily somewhat briefly. There are other items of rather minor importance that operate electrically which, for space limitation reasons, cannot here be dealt with, e.g. cigarette lighters, carburettor throttle control devices, map readers and record changers.

CHAPTER 12

THE WIRING AND INSTALLATION OF AUTOMOBILES

THE wiring system used on automobiles has been much simplified by the adoption of the earth return method, instead of the completely insulated cable system. On the other hand, due to the growth in recent years of the number of electrical equipment items with their individual wiring systems the wiring has become somewhat more complex. However, with the adoption of such expedients as coloured cables, group harness and improved cable connections the wiring installation has been simplified as much as possible. With the earth-return method while it is important to ensure that all of the earth connections are clean and tight at all times, the greatly increased current flow sections of the metal members of the chassis are such that the resistance of the earth return circuit is negligible.

The proper earthing of all electrical components is ensured in special instances by using separate earthing cables, as in the case of the rubber-mounted engine unit which is an important earth return member for the ignition circuits. For this purpose the battery earthing cable is made flexible.

Cables. There are several kinds of cables employed in the wiring of automobiles the conductor sizes or current-carrying capacities of which vary considerably. In general, the selection of the correct area of copper conductor depends largely upon the *voltage drop* rather than on heating considerations.

It is usual to consider the voltage drop of the length of cable used, under its full current loading capacity and to limit this to 10 per cent of the regulated battery voltage, i.e. 1·2 volts for the 12-volt and 2·4 volts for the 24-volt systems.

The cables, in this country, are mostly of the stranded plain copper wire kind in which a number of copper wires are stranded together to provide the required area of copper.

The stranded cables are more flexible than single-conductor cables, but are not so easily soldered. Each size of cable is designated by the number of separate wires, or strands, and the diameter or gauge of wire used. For example, a cable having 14 strands of wire, each of ·012 in. diameter, or 30 S.W.G., is designated a 14/·012 cable.

Types of Automobile Cable. There are three important classes of cables used for the electrical connections of automobiles, namely, cables for (1) The Starting Motor System. (2) General Purposes, e.g. Dynamo, Regulator, Lighting System and other electrical units. (3) The High Tension System.

(1) The Starting System Cable. Since the current at the beginning of the starting motor operation is usually of the order of several hundreds of amperes for short periods, it is necessary to employ a well-insulated cable capable of conducting such currents. There are three British Standard starter cables, each with two alternative insulations, namely, vulcanized rubber and polyvinyl chloride (or P.V.C.) insulation. Particulars of these cables are as follows:

Starting System Cables

No. and Size of Wire	Conductor Nominal Area (sq. in.)	Safe Current Capacity (Amperes)	Average Overall Diameter (inch)
37/·036	0·035	60	0·415
61/·036	0·06	82	0·495*　0·500
61/·044	0·09	175	0·590*　0·600

* P.V.C.

The rubber insulated cable is of the rubber-proofed taped, braided and compounded kind, while the P.V.C. insulated type has P.V.C. insulation, braided and compounded.

(2) General Purposes Cables. For the whole range of automobile wiring requirements twelve different cable sizes are specified by the British Standards Association. These include cables of 9/·012 to 120/·012 single conductor and 9/·012 to 35/·012 for twin conductor cable; there is also a three conductor cable of 9/·012 size.

The most widely used sizes of cables are those for the battery, dynamo, regulator and lamps.

The following are particulars of these cables:

General Purpose Cables

No. and Size of Wire	Conductor Nominal Area (sq. in.)	Safe Current Capacity (Amperes)	Overall Dimensions (Average)	
			Braided (inch)	Armoured (inch)
44/·012	0·0048	22	0·192	0·242
28/·012	0·003	14	0·178	0·207
14/·012	0·0015	7	0·152	0·182

It is necessary when using long cables which produce a voltage drop greater than 10 per cent, to select the next larger size of cable than that given in the specifications.

The insulation for low voltage cables when of the rubber variety should be of the artificial kind, impervious to the action of water, oil or petrol. Neoprene rubber is used largely for this purpose. The American (S.A.E.) cable specifications stipulate thermoplastic insulated braided cables for low tension currents, the cables being of the insulated braidless kind. Thermoplastics are not only stronger and harder than rubbers but are unaffected by exposure to engine bonnet temperatures and to oxygen or ozone of the atmosphere—gases which affect rubber. With the P.V.C. plastic insulation—now superseding rubber—as there is no sulphur content (as with rubber), the conductor wires do not require to be tinned.

Another advantage of thermoplastics is that they are readily extruded and can be made in a variety of colours, with separate tracers.

(3) High Tension Cables. These cables include the cable leading from the H.T. coil to the distributor central connection, to the rotor arm and the cables from the distributor to the sparking plugs. The high voltages, which are of the order of several thousand, carried by the conductors, and their exposure to the engine bonnet temperatures and, occasionally, water, petrol, oil and grease necessitate special insulation requirements. Previously, natural or special rubber insulation was employed, the overall diameter of the cables being from 7 mm. (0·275 in.) to 12 mm. (0·472 in.) and the conductor of stranded tinned copper wire of 35/·012 to 44/·012 sizes. As the currents carried by the H.T. cable are relatively very small the size of conductor chosen is largely governed by considerations of mechanical strength and the necessity of having enough conductor section to make sound strong connections at the ends.

The previously used ordinary rubber cables were affected by heat, oil, petrol and the atmosphere so that after a certain length of service insulation cracks occurred; these occasionally allowed the conductor to short-circuit to adjacent metal parts.

The Neoprene artificial rubber insulated H.T. cable has a marked resistance to the heat, ageing, oil and corona; it has practically replaced other insulating rubbers. The capacitance is much lower for Neoprene than ordinary rubbers.

The S.A.E. requirements for H.T. cables specify a conductor of 7 or 19 strands of annealed tinned copper wire giving a strand group diameter of 0·03 to 0·06 in. Irrespective of the insulation cover over the conductor, a synthetic rubber sheath must be used over the plain insulation or a braided covering, if used. The overall diameter of the cable must be between 0·27 and 0·29 in. Various waterproofing, life cycle, hot oil and temperature tests are also specified.

British Standard H.T. Cables. In 1959[*] the British Standards Institution introduced a third ignition cable and altered the conductor size of rubber-insulated cables, to 19/·012 in. The additional cable is of the P.V.C. insulated type, with plain annealed copper wires, since there is no chemical action between the insulation and copper.

Routine conductor resistance, voltage, ozone, flexibility,

[*] Cables for Vehicles. British Standard 1862:1959.

capacitance and other tests are specified for these cables. The principal particulars of the ignition cables are given in the accompanying table.

Particulars of B.S.I. Ignition Cables

Designation	Insulation	No. and diam. of wires	Approx. diam. conductor	Max. resist. per 1000 yd. at 0° C	Overall diam.
		in.	in.	ohms	in.
7 mm., not braided	Rubber	19/·012	0·060	12·18	0·275
7 mm., braided and lacquered	Rubber	19/·012	0·060	12·18	0·275
7 mm.	PVC	16/·012	0·057	14·19	0·275

Cable Colours. As mentioned earlier automobile cables are coloured for ready identification and to simplify the wiring procedure. In this connection the Society of Motor Manufacturers and Traders (S.M.M.T.) specify *nine* principal colours, which with the addition of colour lines or threads around them, provide a very wide choice of colour combinations.

The S.A.E. specify *fourteen* cable colours, with either black or white stripes, threads or tracers.

The colour code system used with the Lucas wiring system involves *seven* main colours namely, Brown, Yellow, White, Green, Blue Red and Black. Their applications are as follows:

Brown Cables. For the *Battery Circuit*. From the battery or starting motor switch to the ammeter or control box—also to the radio receiver, when fitted—from the control (or regulator) box terminal A1. Also from the starting motor switch to the electric clock, inspection sockets and battery auxiliary fuse (from which are fed the electric horn, cigar lighter, interior lights, etc.).

Yellow Cables. For the *Generator Circuit*. From the dynamo terminal D to the corresponding control box terminal and the ignition warning light.

WIRING AND INSTALLATION OF AUTOMOBILES 351

White Cables. These are used for the *Ignition Circuit* and other circuits used only when the ignition is switched on, but do not require to have fuses, for example, the electric petrol pump motor starter, solenoid switch, etc.

Green Cables. All *Auxiliary Circuits* which are fed through the *Ignition System Switch* but are protected by the ignition auxiliaries fuse, for example, the brake stop lamps, fuel gauge, direction indicators, windscreen wipers, etc.

Blue Cables. *Headlamp Circuits.* These are fed from the terminal S2 (or H) on the lighting control switch.

Red Cables. *The Side and Tail Lamp Circuits.* These are fed from the terminal S1 (or T) on the lighting control switch. Other lamps which are included in this circuit are the fog lamps, panel lights and other lamps which are required only when the side lamps are in use.

Black Cables. *Used for Earth Circuits.* Should an electrical component not be earthed internally to the chassis frame a black cable is taken to a good earthing point on the chassis.

Application to Wiring Diagrams. In the wiring diagrams for modern automobile electrical systems it is customary to indicate the colours of the cables, and any tracers that may be used by various letters or letter combinations, alongside the lines indicating these cables. A key to the colour code is invariably included on, or adjacent to the wiring diagram.

In this connection, Joseph Lucas Ltd. publish a schematic wiring chart showing the circuit relationships of the various items of electrical equipment connected in accordance with the Lucas colour scheme. Various alternative designs of windscreen wipers, electric horns, direction indicators, and overdrives are included in the chart. The simplified wiring method shown in Fig. 223 is being adopted for new models of British cars.

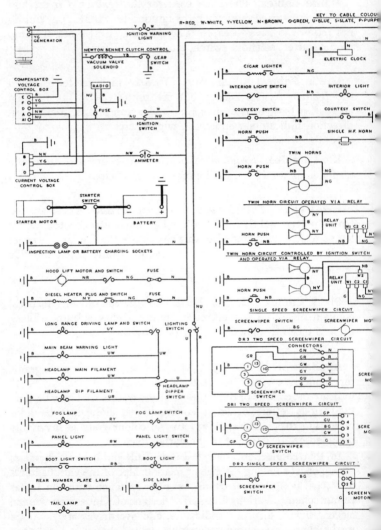

FIG. 223. The Lucas simplified electric

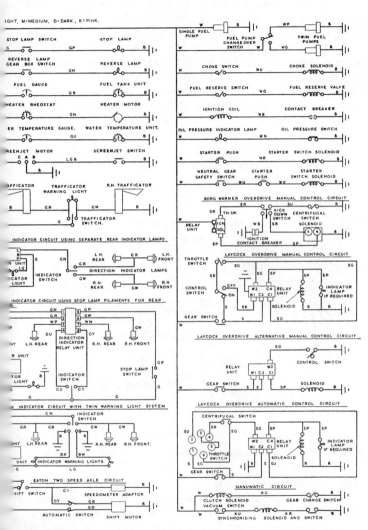

gram, adopted for modern automobiles

Wiring Harness Method. With the relatively complex electrical system of the modern car it would be a lengthy and costly business to wire up each electrical component individually—as was the practice with earlier cars—so that it is now usual to make the wiring into separate groups, known as *harness* or sectional loops which consist of bunches of cables leading to the components to be wired up, in a given vicinity. Each bunch is bound together with varnished tape or P.V.C. tape leaving sufficient lengths of individual cables protruding at each end, to make the necessary electrical connections. The bunching together of the groups of cables enables the wiring of mass-production cars to be carried out expeditiously, while saving a good deal of space and at the same time protecting individual cables against possible damage by abrasion from metal objects.

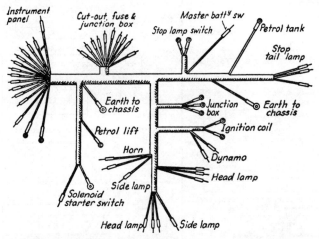

FIG. 224. The cable harness method for automobile wiring

Fig. 224 illustrates the principle of the harness system as applied to a typical car, but without some of the more recent additions to the electrical equipment. The wiring of the instrument panel, at its back face, is often simplified by using a multiple pin plug to which the individual instrument cables are attached

WIRING AND INSTALLATION OF AUTOMOBILES 355

and a fixed socket at the back of the panel, to which the instruments are wired.

While there is the drawback that failure of one cable in the harness would probably necessitate having to cut the harness to replace the cable this is unlikely to happen with modern cables which are of ample mechanical strength and good insulation properties. Should such an eventuality as a broken conductor arise it would be more expedient to fit a new cable externally to the harness and afterwards bound to it.

Harness Cable Connectors. The choice of a suitable cable connector for each of the individual cable ends in the case of the harness system is important, from the time-saving and reliability viewpoints. The earlier method of using spade or eyelet end connectors which are held to the electrical components by screws or nuts, involves a good deal of time, so that some kind of push or snap connector is now employed.

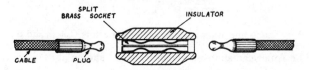

FIG. 225. Lucas quick-action cable connector

An example of a fully-satisfactory and much used cable connector, namely the Lucas design, is shown in Fig. 225. Here a short cable with an insulated socket member—shown on the right—is connected to the terminal of the component by an eyelet or spade connector, while a plug member is attached to the end of the harness cable. The plug is a spring fit in its socket member and can be readily attached or detached.

Some other methods of making cable connectors, used by the same manufacturers, are illustrated in Fig. 226. Of these alternatives that shown at (*a*) gives a more secure attachment, whereas that at (*b*) is quicker for assembly purposes. The fastener shown at (*c*), if the cable end is properly made gives a good spring fit to the anchored socket member.

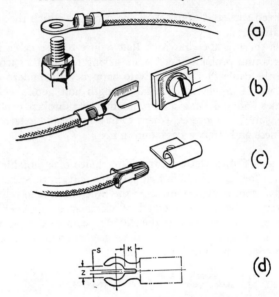

Fig. 226. Some other types of cable connectors (Lucas and S.A.E.)

The American snap-on eyelet terminal shown at (d) is made in a range of sizes to suit terminal screws from about $\frac{1}{8}$ in. to $\frac{3}{8}$ in., the dimensions shown by the letters on the diagram being given in S.A.E. specification tables.

Fuses for Electrical Circuits. In order to protect the electrical equipment against excessive currents fuses—which 'blow' at current strengths below those that would fuse the wires in the equipment—are inserted. In the case of automobiles, owing to the number of circuits and accessories it would be both costly and complicated to insert fuses into each circuit. Therefore, relatively few fuses are employed, so that each fuse has to protect a group of electrical items.

In British mass production cars, most of which use Lucas equipment, it is now the practice to use two main fuses only. One fuse is used to protect the circuits controlled by the ignition switch; its value is usually 35 amps.

WIRING AND INSTALLATION OF AUTOMOBILES 357

The other fuse protects all circuits which are operative whether the ignition switch is 'On' or 'Off'; its value is 35 amps. to 50 amps.

The fuse box is often separate from the dynamo regulator, cutout and cable junction unit and is provided with two spare fuses in a holder on the fitting; the fusing current of the cartridge type fuse used for automobile purposes is marked on a coloured paper slip inside the glass cover of the fuse.

The fuses used on commercial vehicles are usually of the strip kind, although fuse wires are also used. Strip fuses of special fusing metal are of flat form with a hole at one end and a hook at the other to connect between two screw terminals.

American Practice. The electrical circuits are protected against the effects of excessive currents by either of the following two methods, namely, (1) *Fuses* and (2) *Circuit Breakers*.

(1) *Fuses*. Individual component circuits or those of groups of components are usually protected by separate fuses. Examples of these components or accessories, with their fuse values, in brackets, include some or all of the following, in a typical American car: Air Conditioner and Heater (25 amp.), Interior Lamps (7·5 amp.), Radio (7·5 amp.), Radio Antenna Motor (14·0 amp.), Cigar Lighter (15 amp.), Reverse Lights (7·5 amp.), Overdrive

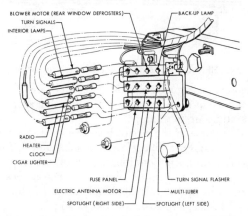

FIG. 227. Fuse block of typical American car (Mercury)

(15 amp.), Direction Indicators (7·5 amp.), Dynamo-Battery Circuit (40 amp.) or according to maximum charging rate. The following group is protected by a 25 amp. fuse: Windscreen Wiper, Clock, Interior Lights, Map Panel, Indicator Lights.

It may be mentioned that occasionally a circuit breaker is used for some of the components mentioned, instead of a fuse. Usually, the fuses are located in a single fuse block, as shown in Fig. 227, but the dynamo fuse is separately mounted, as a rule.

Circuit Breakers. These units which may be regarded as non-replaceable substitutes for fuses each consist of a bimetallic strip device having a fixed and moving contact, the latter being at the free end of the strip. When the current through the unit exceeds the maximum permissible value for the electrical component concerned the strip bends so that its contact moves away from the fixed contact thus breaking the circuit. When used in a lighting circuit the lamp will light and then go out, repeatedly, thus calling attention to the faulty circuit (Fig. 228).

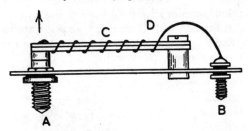

Fig. 228. Showing principle of the bimetal circuit breaker

The circuit breaker unit (Fig. 228) has a pair of screwed terminals A and B, a current heating coil C and bimetallic strip unit D. It is made in ratings up to 50 amps.

A typical example of the circuit breakers used in an American car is, as follows: For: The Headlamps (18 amp.), Low Beam Headlamp (12 amp.), Air Conditioning Motor (20 amp.), Electric Window Operation (14 amp.), Four-way Power Seat Motors (20 amp. and 30 amp.). Group of the following lights: Parking, Rear, Stop, Licence, and Instruments (12 amp.).

CHAPTER 13

LATER DEVELOPMENTS IN ELECTRICAL EQUIPMENT

SINCE the last edition of this Manual was published there has been a number of new developments in the electrical equipment for cars and other light vehicles, although many of these embody the basic principles dealt with previously.

The most important improvements are those based upon the applications of semi-conductors, printed electrical circuits and in some instances, microelectronics or miniaturization of electrical printed circuits.

Of the semi-conductors, transistors and diodes are the principal ones used in automobile components. In this connection their applications include the rectification of A.C. currents, miniature voltage regulators, high-speed ignition systems, flashing-type indicators, anti-dazzle headlamp beam controls, radio, car clocks, electronic speedometers and tachometers.

Transistors. These depend upon the use of semi-conductors, which materials are neither good electrical conductors nor good insulators. They also differ from the usual conductors in that their electrical resistance actually diminishes with increasing temperature. Thus, in the graph shown in Fig. 229, the temperature-resistance for all ordinary conductors is represented by the line AB, showing the linear increase in the resistance with temperature. The graph for a semi-conductor has the general curve shape CD, showing a fall in resistance with temperature increase.

Of the available materials now largely used for semi-conductors, silicon and germanium are the more important, the former being used for modern transistors and the latter for diodes.

Although an earlier application of semi-conductors is referred to on pages 109 to 113, it is here proposed to give some general information on these, from the atomic structure angle.

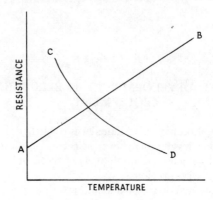

Fig. 229. Temperature-resistance graphs for electrical conductor *AB* and semiconductor *CD*

Atomic Structure Considerations. The atom consists of a central member, or core, containing protons surrounded by electrons which for explanatory purposes are here regarded as concentric to the core; in fact the orbits are elliptical and inclined. The proton is the electrically positively charged member and the electron the negatively charged one.

Metals, e.g. copper and aluminium, which are good conductors, each have four electron orbits, but contain in their outer orbit less than four electrons. Materials which are good electrical insulators have more than four electrons in the outer orbit.

In the case of the element germanium this has four orbits consisting from the centre outwards of 2, 8, 18 and 4 electrons. By the addition of an impurity, such as arsenic or antimony (each of which has five electrons in its outer orbit) a certain effect, known as co-valent bonding, occurs. This results in the existence of a free electron which can pass freely through germanium-bond material, if an appropriate E.M.F. is applied. The germanium-bond material is known as a negative, or N-type material.

Again, if an impurity such as the element indium which has three electrons in its outer orbit is added to germanium, the result is equivalent to an empty space, known as a 'hole' and having a positive charge, being free to move in the material, which is then known as a P-type one. Another important element, namely silicon, has three orbits with 2, 8 and 4 electrons, respectively. In some applications it is superior to the germanium-bond material, e.g. for alternating current rectifiers.

Diodes. When flat sections of N-type and P-type materials are joined together the combination, Fig. 230 (A), is known as a diode, the symbol for which is shown in Fig. 230 (B). The arrangement of a typical diode is shown schematically at (C), in Fig. 230. The conductor A, which is mounted in a cylindrical casing D, is attached to a thin plate of silicon C, by a flexible member B. The casing D apart from protecting the diode member also acts as a heat conductor, or 'heat sink', which dissipates any heat generated during the operation of the diode.

Fig. 231 shows an electrical circuit containing a source of alternating current supply at AC, the diode D and resistor R. When the positive loop of the alternating current occurs, as shown in Fig. 231 (a), this is known as 'forward bias', and direct current, but of varying strength, will flow in the direction of the arrows. However, during the negative loop of the alternating current cycle (Fig. 231 (b)) the diode will receive what is known as a 'negative bias' since the voltage at AC is then negative so that in this case no current will flow in the circuit. It will be seen that the diode allows only the positive part of the A.C. to flow—a condition known as half-wave rectification. As explained later, modern alternators use circuits giving much greater outputs than the one we have described.

Transistors. The transistor consists of three joined-elements, either of the PNP or NPN types. In the former case a P-type element is connected to a PN diode, to give the arrangement shown at F, in Fig. 232, in which a plate of N-type material is connected to short cylinders of P-type material, with the conductor wires at c. The electrical symbol for this transistor is given at H and the names of the three conductor items, namely Emitter, Base and Collector are denoted by the letters E, B and C on the diagram. The arrow denotes the direction of current flow. In the transistor shown at (G) the plate is of P-type material, with sections of

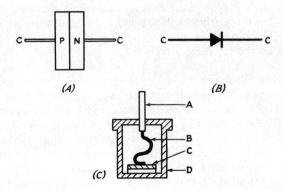

Fig. 230. The silicon diode. (A) P-type and N-type materials. c—conductor (B) symbol for a diode. (C) a diode unit: A—conductor, B—flexible conductor, C—silicon disc, D—casing

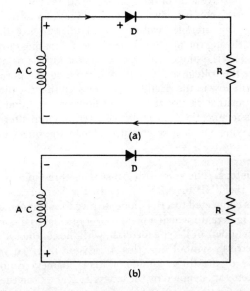

Fig. 231. Operation of a diode. (a) forward bias rectifying circuit. (b) reverse bias non-rectifying circuit

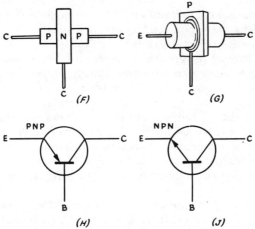

Fig. 232. Transistor units and symbols. E—emitter. C—collector. B—base

N-type material on either side of it, as shown by the symbol at *J* in Fig. 232. The arrow shows the current flow direction.

It has been possible to give only a very brief outline of the principles of diodes and transistors, but for those desiring fuller information the references at the end of this Chapter should be consulted.

Notes on Transistors. Some examples of electrical circuits having diodes and transistors are given in this Chapter, in which the methods of connecting these units are illustrated.

If the voltage of the current supply to a transistor is a certain limiting value the current will increase above its correct limit and the element will generate more heat, thus causing a reduction in the resistance and a further increase in heat, finally giving rise to a condition known as 'Thermal runaway' which can damage or destroy the transistor. It is necessary to arrange the circuitry so that this effect cannot occur. In some cases a voltage limiting device known as a Zener is introduced into the circuit. It is of the reversed bias silicon diode type, and is often used as a voltage stabilizer in circuitry.

The provision of heat sinks is important in diode and transistor applications; these heat conductors and dissipators help to avoid the temperature effects on their resistance.

Transistors of the germanium type, according to their dimensions, have limiting temperatures, usually between about 70° C and 80° C. In the case of silicon transistors a much higher temperature, namely, about 135 to 145° C can be employed.

The transistor is capable of most of the duties of the thermionic valve used in radio and television sets and has indeed supplanted the valves in recent sets. It can act as a detector, amplifier and oscillator with the marked advantages of much smaller bulk, lighter weight and greater efficiency over a wider range of electrical outputs.

In its modern applications the transistor has the following other advantages, namely: (1) Increased reliability. (2) Is unaffected by shock and vibrations within fairly wide limits. (3) Has a relatively long life: thus in some automobile applications will last at least 100,000 miles. (4) Has no moving parts. (5) As previously stated it enables components to be made very much smaller. (6) Is unaffected by moisture. (7) Can readily be used in micro-circuits. (8) Can produce a relatively large current output from a small input one.

Printed Circuits. These are now widely used in electronics, a typical example being that of computers. More recently the printed circuit has been employed in certain automobile units, e.g. alternator output controls, two-level illumination systems, direction indicators, and instrument panels.

The principle of the printed circuit is the use of an etched or stamped-out copper matrix with conducting strips to take the place of wires, and the embodiment or attachment of certain electrical components, such as capacitors, resistors and transistors. Suitable terminal holes for outside cable connections are included. In a typical example a flat and usually thin sheet of insulating laminate, e.g. epoxy-resin, glass fabric or resin-bonded paper is given an adhesive-coated layer of copper sheet, of suitable thickness on one or both sides. The required circuit is made as a black line on white drawing-paper which is photographed down to the required size. The copper plate is coated with a special photo-sensitive 'resist' and when a negative from the photographed circuit is placed emulsion side down on it and exposed to light through the outer glass side, the effect on the dense parts of the plate is to make the resist soluble in a special developer, while the transparent parts of the negative, corresponding to the circuit black lines are fully resistant to the same developer. The result is that the developer dissolves the resist on the light exposed regions, leaving the resist over the circuit line and other protected areas. The copper plate is then washed and placed in another solution which dissolves the copper right away down to the laminate base, leaving the circuit matrix with its resist still intact; the latter is then removed in another chemical solution.

Yet another method involves the use of press tools with a heated male die to represent the circuit required. The copper circuit is pressed out from a plate on to an adhesive-coated laminate base; this method enables thicker copper circuits to be made.

FIG. 233. Rear view of automobile panel printed circuit

Fig. 233 shows the rear view of the Ford Corsair printed circuit instrument panel made by S. Smith & Sons Ltd. The panel incorporates the voltage regulator, two instruments—for water temperature and fuel level, five warning lights—for direction indicators, high headlamp beam, ignition and oil pressure warning lights and two internal light bulbs. An 11-pin connector connects the panel to the wiring harness by a single operation.

Flexible Printed Circuits. Instead of using flat sheet panels for copper circuits these can now be made flexible, so as to conform to automobile instrument and control panel requirements. An example of this type is the Lucas Flexprint on which the printed circuitry is permanently bonded between two flexible films of insulating plastics, so that it can be bent, shaped, rolled or folded.

Both types of printed circuit panels greatly simplify the instrument and control wiring on the complete panel, thus avoiding wiring errors and maintenance of the various permanent connections to the panel.

The Negative Earth Method. Hitherto, for the reasons given on pages 29 and 30, the positive earthing system was adopted for English cars, whereas the negative earth one was the standard system in the U.S.A. and certain other countries. More recently, however, certain alterations have been made in the designs of components, e.g. the battery and ignition system items, while the alternator has replaced the dynamo in later cars, so that it has become the practice to use the negative earthing method in this country, also. It was mainly the replacement of the dynamo by the alternator with its semi-conductor non-vibrating voltage control which influenced the adoption of the negative earthing system.

The problem of battery terminal corrosion has largely been overcome in more recent batteries by design and metal alloy improvements. In the case of the ignition system the advantages of the positive earthing system in regard to sparking plug electrode erosion and also that of the metal end of the distributor rotor have been overcome by reversing the low tension connections to the SW and CB terminals, from the positive earthing system, when the negative one is adopted; the same H.T. coil is used in either system.

Alternator Developments. Certain important developments have been made since the models of these A.C. generators were described in the last edition of this Manual. Alternators have become almost standard practice in the U.S.A. and there are now various British, European and Japanese makes fitted in modern automobiles.

Since the earlier introduction of these alternators, notably by Lucas and AC–Delco, in this country, the overall dimensions and weights have been progressively reduced. The accompanying Table covers the Lucas alternator improvements, since 1930, and shows the notable reductions in weight—and therefore dimensions —and increases in output over the period mentioned.

LATER DEVELOPMENTS IN ELECTRICAL EQUIPMENT 367

Developments in Automobile Alternators

Date	1930	1940	1955	1960	1965	1968
Ampere rating	12	12	19	30	35	28
Weight in lb.	24	16	12·5	17	9	7·5
Output, watts per lb. weight	7	10·5	21	24	52	53

The more favoured arrangement for modern alternators is that of a three-phase stator output winding with six diode rectifiers of the silicon type, as shown in Fig. 234, for a battery charger—but without the usual voltage regulator. The windings are made on a ring-shaped lamination pack, with a multi-pole rotor carrying a pair of

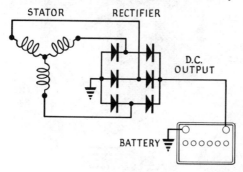

FIG. 234. Typical arrangement of alternator stator windings and diodes (Delco–Remy)

slip rings. Two long-life positive and negative brushes take their current supply from the slip rings. In the Lucas 10AC and 15AC alternators the slip rings are in the form of concentric rings, with their brushes bearing sideways on to their surfaces. In regard to the diodes, these are mounted on the end cover of the alternator and are provided with heat sinks; they are cooled by a fan-induced air stream through the space between the rotor and alternator casing.

Output Current and Voltage. No current regulator is necessary with the alternator, since it has self-regulating characteristics, but in regard to the voltage, this must have a regulator to keep it within the usual prescribed limits over the speed range. The previous transistorized type of vibrating blade regulator, referred to on page 118, has more recently been replaced by a new type of relatively small size, which, using transistors and diodes, provides more accurate control of the voltage and can operate with a large field current to give improved performance. As this type of regulator

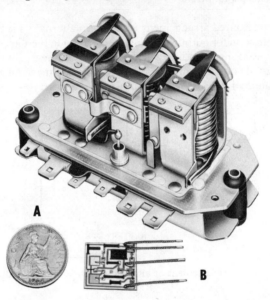

FIG. 235. Comparison of Lucas micro-circuit unit B, with a penny A and the original current–voltage regulator which it displaces in the modern alternator

has no moving parts it requires no maintenance. In the recent Lucas alternators this voltage regulator consists of a thick-film integrated micro-circuit having resistors and palladium-silver conductors, screen printed on to an alumina substrate with two capacitors and five semi-conductors mounted as discrete components. Fig. 235 shows this miniature voltage regulator, in comparison with a penny coin and the original current-voltage regulator above. The small regulator is housed in an aluminium heat sink sealed with a silicon rubber encapsulant.

LATER DEVELOPMENTS IN ELECTRICAL EQUIPMENT 369

Alternator Circuit Warning Lamp. In the previously-used car dynamo a red warning lamp was connected across the cut-out contacts to give a warning when the engine was stopped or idling slowly, with the ignition switch 'On'. With the alternator it is not satisfactory to connect the field circuit through the ignition switch due to the existence of varying loads on the switch which would give rise to voltage variations across it, thus interfering with the alternator voltage control system; also, there is no need for a cut-out, as explained earlier. There are, however, alternative methods for illuminating a warning lamp, one of which is to use a double-pole switch, one pole of which cuts out the alternator field circuit while the other switches 'Off' a warning lamp in a thermal relay circuit. In this case, alternator rectified voltage is used to put out the warning lamp and also to supply the control system (Fig. 236).

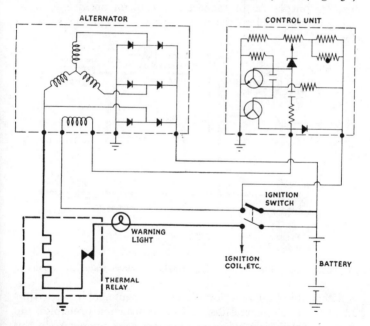

FIG. 236. Showing use of a double-pole ignition switch to isolate the alternator field for the ignition warning light

Another method, shown in Fig. 241,* employs three more diodes in the alternator circuit to give rectified current at battery voltage. In all instances the warning light is extinguished when the alternator is operating.

The layout of the principal components of an AC–Delco alternator, having a transistorized voltage regulator, relay and indicator or warning lamp, is shown in Fig. 237. In this system the alternator output is taken from the BAT terminal on the alternator (generator) to a junction box, or block, from which the various automobile items are fed. The field terminals F-1 and F-2 are connected to the voltage regulator, shown on the right, from the positive terminal of which the ignition switch, indicator lamp and relay are connected in series to the junction box. The field relay is connected to the regulator, one lamp terminal and the junction box. Its purpose is to connect the regulator positive terminal and generator field to the battery when the ignition switch is closed.

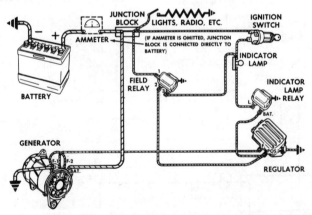

FIG. 237. Typical American charging layout with transistor regulator

Alternator Construction Notes. Most automobile alternators adopt the same general design and construction principles, but differ in detail, design and arrangement of components. In this

* *Vide* Reference No. 2 at end of this Chapter.

connection, an example of an alternator embodying the experience and development lessons over a long period is that of the Lucas 10AC alternator. This is one of two models with similar designs but different outputs, as shown in Fig. 238. While these alternators are capable of safe operating speeds up to about 12,500 r.p.m., their maximum regulated speed is 13·5 volts at 3,000 r.p.m. The automobile dynamo commences generating current at about 700 r.p.m.

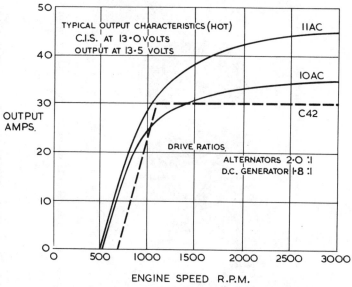

FIG. 238. Comparative alternator and D.C. dynamo performance graphs

whereas the alternator, operating at about 700 to 800 r.p.m. will give a charging current of 10 to 13 amps. At the dynamo's cutting-in speed of 1,000 to 1,100 r.p.m. the alternator gives a much higher output. The dynamo equivalent to the 10AC alternator had a maximum output of 30 amps. and weighs 17 lb.

Fig. 239 shows the components of the 10AC and 11AC machines in their respective positions along the common axis. These include the imbricated pole-type rotor with its concentric side-facing slip rings, the stator with its laminations, the diodes with their heat

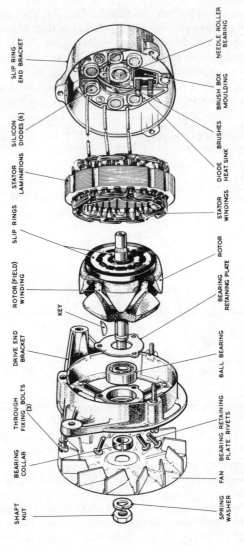

Fig. 239. Components of Lucas 10AC and 11AC alternator.

LATER DEVELOPMENTS IN ELECTRICAL EQUIPMENT 373

sinks in the end cover, the other driving side end cover, and cooling fan for the interior parts of the alternator. The stator comprises a 34-slot, three-phase star-connected output winding on a ring-shaped lamination pack, housed between the slip ring end and cover and the drive end bracket. The rotor is of the 8-pole kind and carries a slip ring field winding. It is supported by a ball-bearing at the drive end and a needle roller bearing at the other, or slip ring end cover. The brush gear for the field system is mounted on the latter end cover. The negative and positive carbon brushes bear against the concentric slip rings, the positive brush being on the inner slip ring, to reduce its known higher rate of wear. The slip ring cover also carries the six silicon diodes connected in a three-phase bridge circuit for rectification purposes. The diodes and stator windings are cooled by the 6-in. diameter fan at the drive end. The alternators have Model 4TR regulators each controlling the voltage to 13·5. These regulators are of the vibrating contact type with parallel field resistor. It includes the semi-conductor diode system mentioned earlier.

FIG. 240. The Lucas 15ACR alternator, with end cover removed.

Alternator with Inbuilt Transistorized Voltage Regulator.
The Lucas types 15ACR and 16ACR alternators designed for new cars or as replacements for dynamos are of similar designs but different outputs. While intended for negative earth systems, existing positive earth system cars can readily be modified to take these alternators. The alternator, shown in Fig. 240 commences generating at 1,000 r.p.m., generator speed. At speeds of 2,000, 4,000 and 6,000 r.p.m. the respective current outputs are 16, 25 and 28 amps. The voltages are between 13·5 and 14·4 at 20° C ambient temperature. The safe operating temperature range is between minus 30° C and 93° C.

Fig. 241 shows the 15ACR circuit diagram, for the alternator with its Type 8TR regulator. It will be seen that the alternator output is rectified by six silicon diodes which are housed in a rectifier pack and connected as a three-phase full wave bridge. The rectifier pack also contains the three field diodes through which, at normal operating speeds, rectified current from the stator output windings flows to provide self-excitation of the rotor field, *via* the brushes on the slip rings.

It may here be mentioned that provision is made for a radio suppressor capacitor to be fitted on the slip ring end bracket. The alternator has the small micro-circuit voltage regulator shown in Fig. 235, from which connections to external circuits are brought out to special connector blades, while the main output terminals are arranged to take a moulded connector socket on the car's cable harness. In operation the action is similar to that of an electromagnetic-type regulator in that the current in the alternator field winding is varied to maintain the output voltage within close limits, but the control is achieved by a Zener diode Z and three transistors, instead of the previous vibrating blade contacts.

The Ignition System. Although the conventional system, with its H.T. coil and contact breaker unit has given good service over a long period, more recently its design and construction have been improved to give even better performance, improved reliability, longer useful life and extended maintenance periods. In particular, centrifugal and vacuum advance mechanisms and layouts have been changed, while the contact breaker unit with its relatively

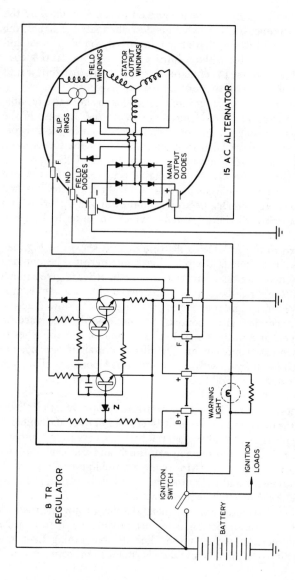

Fig. 241. Circuit diagram for 15ACR alternator fitted with 8TR regulator battery charging system

large number of parts has been replaced in some engines by the single unit contact breaker, which is described later. Further, the distributor shaft bearings have been improved, to reduce wear and give prolonged service without lubrication attention. In this connection it is of interest to compare the more recent distributor unit shown in Fig. 242 with that of Fig. 101, on page 170.

While the conventional ignition system has been in use on production car engines over a long period, with the increase in the maximum operating speeds of some newer models, transistor-assisted systems have been introduced in either kit forms or as alternatives to the conventional system. In the U.S.A. certain makes of car engine can be purchased with either of the two alternative systems mentioned. Apart from the transistor-assisted systems more attention has been given to the fully transistorized and also the capacitor discharge systems, which are described later in this Chapter.

In regard to the AC–Delco distributor unit shown in Fig. 242 among its improvements, the following are of interest. The moulded cap is provided with heavy internal ribbing in order to give a much longer 'tracking' resistance path. The contact breaker cam is profiled to give more rapid opening to the contacts with slower closing to prevent contact arm 'bounce' at high speed. The vacuum advance mechanism breaker plate is mounted on nylon bushes to give a much easier movement. The distributor shaft bearings are made of sintered iron, with finely controlled porosity. They are kept lubricated by the oil bath above the upper bearing.

Contact Breakers. These have been much improved, the following being the more important changes: (1) Larger tungsten contacts. (2) The replacement of fibre material by harder wearing synthetic ones, e.g. nylon for the moving arm bush and the cam-rubbing heel on the moving arm. (3) More compact and lighter design, with fewer components and therefore greater reliability and (4) Introduction of the single 'Quikafit' contact unit.

In a later distributor unit the metal diaphragm for the vacuum control of the ignition timing, is replaced by one made from a Du Pont elastomer, known as Viton. The flexible tubing from the vacuum unit to the inlet manifold is also made in the same material.

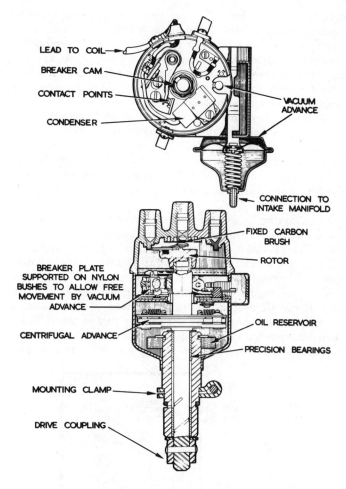

Fig. 242. Plan and side-sectional view of modern contact breaker and distributor units (AC–Delco)

378   MODERN ELECTRICAL EQUIPMENT

In the distributor unit there are two relatively small vacuum advance units, instead of the usual single one.

The previous design of the Lucas contact breaker assembly consisted of nine separate components, as shown at L in Fig. 243. This design has now been replaced by a single unit S, on the left, known as the Quikafit, which contains both the fixed and moving arm contacts and a stainless steel spring S. The heel of the moving arm is made from plastic material, Kemetal, which has much greater wear resistance than hitherto. The contacts are pre-aligned, so that no adjustment is required when replacing the previous assembly. The moving arm is of low inertia, to reduce or eliminate 'bounce' effects at high engine speeds. The contact points are of a self-cleaning nature, giving much longer maintenance periods.

Cold Starting Ignition System. Under severe cold weather conditions engine starting is more difficult than in normal circumstances. This is due to (1) The greater starting torque needed due to the more viscous engine oil and (2) The fact that the battery capacity is lowered by the cold temperature.

To render engine starting much easier, in cold weather, the Delco–Remy 'Cold Start' system was devised. It has a special 2-ohm ballast resistor attached to the upper part of the H.T. coil which is designed to operate at about 7·5 volts for a 12-volt battery. The resistor is fitted in the low tension circuit for this purpose.

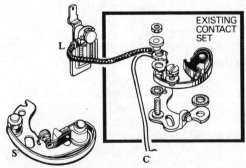

FIG. 243. Single unit contact breaker with, on right, the previous multi-component one

Normally, the ignition system operates on this lower voltage, but when the engine is to be started a relay causes the resistor to be by-passed so that the full battery voltage of 12 is applied to the coil, to give a much higher voltage spark at the plugs. The by-passing is done during the cranking action, by the use of a 'finger' in the solenoid starter switch which is connected to the coil primary circuit, thus giving the full battery voltage. Fig. 244 shows the circuit diagram for this ignition system. This method of starting reduces battery drain in cold weather and avoids excessive starting

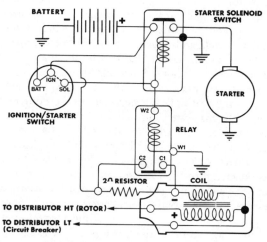

FIG. 244. Circuit diagram for special cold-starting system

motor wear. The system, however, can only be used on an automobile in which the ignition key not only switches on the ignition but also, on a further 'turn' operates the starting switch. It is used on certain American cars and more recently on English vehicles.

In Fig. 244 the relay terminals W1, W2, C1 and C2 are connected, respectively to Earth, the Solenoid of the Starter, the H.T. Coil Primary and the Ignition side of the Resistor.

Transistor-assisted Ignition Systems. These systems, mentioned earlier in Chapter 5, employ the conventional contact breaker and distributor units but provide transistors to produce higher voltage sparks at the sparking plugs. The advantages of the later systems, may be summarized, as follows: (1) Reduced non-inductive current at the contact breaker points. (2) Longer operation period, e.g. at least 25,000 miles for the plugs, instead of the usual 10,000 miles. (3) Improved engine performance. (4) Better engine starting from cold. (5) No need for the conventional contact breaker condenser. As an example of a more recent system, the Lucas T.A.C. has been chosen. This employs a special high voltage type transistor to give very good switching characteristics, since it actually makes and breaks the primary circuit current, thus inducing high voltage sparks at the plugs. Its silicon transistor can withstand 500 volts and interrupt a current of 5 amps. continuously. The maximum power that can be dissipated is about 50 watts. The transistor is housed in a small metal case, having also a capacitor and a printed circuit with resistor and an aluminium-base to act as the heat sink. It is necessary to use a different type of H.T. coil with this system, namely one with a reduced primary inductance to that of the conventional coil.

Fig. 245 shows the circuit diagram for the T.A.C. ignition system with its various components. The transmitter T has its base b connected through a resistor $R1$ to the contact breaker CB terminal. The switching action of the transistor occurs when the contacts close; this allows the much larger current to flow through the primary coil circuit by way of the collector and emitter electrodes, shown at c and e on the transistor. When the contacts are opened by their cam the electronic switch becomes non-conducting, so that the flux in the core of the coil collapses, thus causing the secondary winding high voltage to produce a spark at the plug. Since the transistor is capable of switching larger currents than for the conventional ignition system, the primary coil inductance is reduced, as stated earlier. The ballast resistor shown in series with the primary coil is to compensate for the low resistance of the coil winding, to reduce internal heating.

It may be mentioned that conversion kits are now available, from some electrical manufacturers to enable one to convert from the conventional to the transistor-assisted system. These kits include a new H.T. coil, the transistor pack and a ballast resistor.

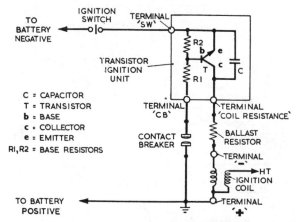

Fig. 245. Circuit diagram for Lucas T.A.C. transistor-assisted ignition system

American Ford Car Transistor-assisted Ignition System.
Fig. 246 shows the circuit diagram for this system, with the various components used in their respective positions. It is stated that the system gives a current reduction through the contact breaker points from a maximum of 5 amps. to 1 amp. It increases the primary coil current and gives a sparking plug voltage which is almost independent of engine speed over the whole range. Referring to Fig. 246 the electrical system shown includes the two-position ignition key switch and the starting motor switch; the lower or second position of the key switch starts the engine. From this switch current flows through a ballast resistor to the transistor pack which has connections to the contact breaker points and to another ballast resistor in series with the primary of the special-type H.T. coil. The secondary discharge goes, *via* the distributor (not shown) and thence to the plugs. Reverting to the transistor unit this includes the transistor, two capacitors and a Zener diode which not only limits the forward voltage across a transformer—below the transistor—but also dampens the reverse voltage across the transistor when the primary current tends to a higher voltage than that of the battery.

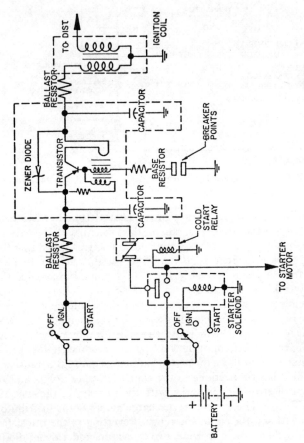

Fig. 246. American Ford car transistor-assisted ignition system circuit

The results of tests on this system showed that at 1,000 and 4,800 r.p.m. the secondary circuit voltages were 26·8 and 22·8 kilovolts, whereas for the conventional system they were 25·3 and 10·7 kilovolts, respectively.

Fully Transistorized Ignition Systems. In these systems no contact breaker is used, so that there are no longer the maximum speed limitations enforced by that system. This fact is illustrated in Fig. 247 which gives a comparison between the performances of the coil ignition and the Lucas transistorized system. It shows that the practical upper limit for the coil system is about 400 sparks per second, whereas with the transistorized or electronic one it is possible to obtain a constant at least 1,000 sparks per second, at a constant output of about 22·5 kilovolts. It can be shown that a rate of 1,000 sparks per second will supply the needs of an eight-cylinder engine operating at 15,000 r.p.m.

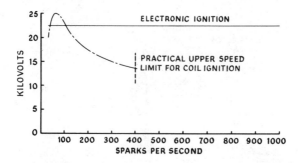

FIG. 247. Sparking performances of the conventional and Lucas transistor-assisted ignition systems

In this system the ignition timing is determined by an electromagnetic triggering device. The principle of the transistorized system is illustrated in Fig. 248 in which the positive pole of the battery B is connected through the switch *SW* and resistor *R* to a control unit *C* containing the transistor and its other components, which act as an electronic switch. This is actually switched 'On' and 'Off' by a triggering device—in this case a mechanically-driven magnetic pulse generator—denoted by T. The switching action of the unit C produces a short duration large current in the primary coil circuit P, which in turn gives a high voltage current in the secondarys of the H.T. coils H which goes *via* D to the distributor unit.

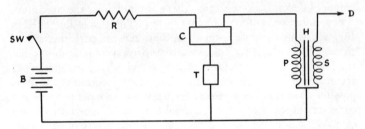

FIG. 248. Illustrating the principle of the fully transistorized ignition system

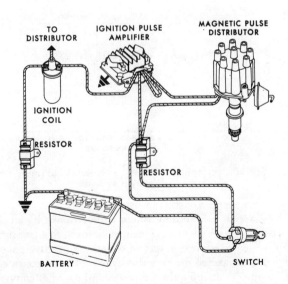

FIG. 249. Layout of typical American transistorized ignition system

The Delco–Remy system has four units, namely, a magnetic pulse distributor, a control unit with three transistors, a special H.T. coil and a resistor. Fig. 249 shows the arrangements of the components and their connections. The magnetic pulse generator, Fig. 250, resembles the usual coil-ignition type distributor unit, but inside is a magnetic pick-up assembly consisting of a bearing plate on which are sandwiched together a ceramic ring-type permanent magnet, two poles and a pick-up coil. The pole pieces are doughnut-shaped steel plates with accurately-spaced internal teeth, namely, one tooth for each engine cylinder. The timer core

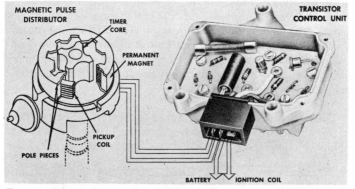

FIG. 250. The magnetic pulse generator and transistor control unit of the Delco–Remy ignition system for 8-cylinder engine

has a number of equally-spaced projections or vanes and is driven by the usual distributor shaft at half engine speed. As the vanes of the iron time core pass near the pole teeth the magnetic field alternately builds up and collapses, so that a voltage pulse is induced in the pick-up coil, each of these pulses being conducted to the transistor control unit where it turns 'On' the triggering transistor, causing it to turn 'Off' the switching transistor. This action stops the current flow through the H.T. coil primary winding thus producing the high voltage spark at the sparking plug. The switching transistor then automatically returns to an 'On' condition permitting the coil current to build up for the next firing operation. It may be mentioned that the transistor control unit, has no moving parts. It contains three transistors, a Zener diode, a condenser and five small resistors.

The Lucas OPUS Transistor System. This contains an electronic oscillator which is located above and driven by the usual distributor drive shaft of the conventional coil ignition system, at one-half engine speed. The shaft drives a small drum having as many ferrite rods as the engine has cylinders. A static unit which surrounds the rotating drum has a specially-shaped ferrite core provided with limbs on each of which is a coil winding. These static windings are all connected in series, but are arranged across the output terminal of the rotating member, or oscillator. The pulses generated by this combined unit are fed to a transistorized amplifier unit circuit which switches off a power transistor controlling the primary coil current. The switching off of the primary current causes a high voltage discharge from the secondary coil across the sparking plug electrodes.

The Capacitor Discharge Ignition System. The principle of this alternative system, which is particularly suited to the needs of higher engine speed production cars, is shown in Fig. 251. From

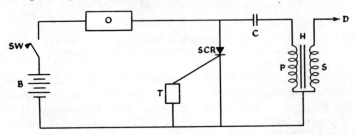

Fig. 251. Illustrating the principle of the capacitor discharge ignition system

the positive pole of the battery B current flows through the closed switch SW, to an oscillator O which transforms battery voltage to alternating, or pulsating voltage which is rectified, by a silicon diode rectifier, SCR to a steady voltage of a few hundred, which then charges the condenser or capacitor C. The triggering of the rectifier or transistors in the case of transistorized switching systems in order to control the primary current, is effected by the trigger device shown at T; this can be either a conventional contact breaker, or a magnetic pulse generating device. When triggering occurs the condenser is discharged through the primary winding P, of the H.T. coil, shown at H, thus causing a high voltage discharge through the secondary circuit S and through the distributor at D to the plug. In this system the secondary voltage has a more rapid rate of increase and a shorter duration than in the transistor and conventional contact breaker systems.

In a typical capacitor discharge system the pulses range from about 0·25 to 35 volts and their operating rate can reach about 800 per second. An advantage of this method is that there is a good spark at the plug at low engine speeds and even when the engine is motored for starting from the cold, or when the battery is not well charged. This system can be designed to give a 30-kilovolt spark at the plug, over the engine speed range. In regard to the duration of the sparks, in some comparison tests concerning plug fouling causes the capacitor discharge system gave 15 kV. micro sec., while the conventional ignition system gave a sparking voltage rise of 2 Kv. micro sec.

The Sparking Plug. While the types of plugs previously used have given satisfactory service and useful lives of at least 10,000 miles for their engines, later engines of the overhead camshaft kind, using higher compressions and limiting valve head sizes to obtain better volumetric efficiencies, necessitate smaller combustion chambers with less room for plug accommodation, taking the cooling of the plug into account. Further, it has been found that after a car has been run at high speeds on a motorway, followed by slow traffic speeds in cities there is an increasing tendency for the plugs to foul. Also, with modern higher output engines a somewhat different design of plug is required. In this connection the colder type of plug, as shown in the half-sectional view at A in Fig. 252 has a longer clear insulator nose, and gives a better heat flow path for cooling the plug than the one shown at B, which is a hot-type plug that is recommended for normal speed and cooler-running engines. Both plugs indicated in Fig. 252 are of the 14-mm. kind

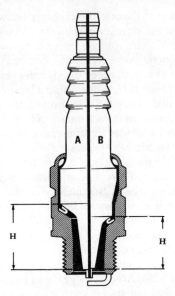

FIG. 252. The AC sparking plug types. (A) cold-type. (B) hot-type

and have four-groove top insulators, to minimize flash-over from the terminal to the metal shell below. Each plug has a fused copper and glass centre conductor giving a gas-tight seal. The sparking electrodes are of larger dimensions than hitherto, to prolong the useful life of the plug and also to give easier starting of the engine. The thin tip insulators of both types of plug resist fouling and pre-ignition, the temperature of the tip following that of the combustion chamber to some extent, heating up fast to burn off deposits and cooling more readily to prevent pre-ignition. The plug described is known as the AC Hot Tip one.

Plug Gap Positions in the Combustion Chamber. The location of the plug gap in relation to the surface of the combustion chamber is an important factor in engine operation. Fig. 253 shows three different types of sparking plug ends which have been carefully investigated.* Of these, that shown at *A* was designed to give

* See Reference No. 5 at end of this Chapter.

LATER DEVELOPMENTS IN ELECTRICAL EQUIPMENT 389

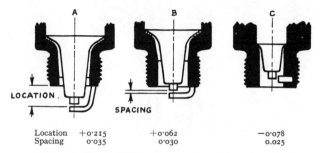

	A	B	C
Location	+0·215	+0·062	−0·078
Spacing	0·035	0·030	0.025

FIG. 253. Three different types of sparking plug tips

the wider heat range that was needed for modern overhead valve engines. In this design the protruding portion is kept cool by the high velocity incoming mixture while at small throttle openings the temperature increases so as to burn off any fouling deposits. It was also found that the plug *B* gave a lower fuel consumption than the other two. This type of plug is most suitable when cars are driven in large towns and cities, after coming from motorways. The plug shown at *C*, for racing car engines, has the shortest earth electrode, in order to keep its temperature down.

Ignition System Effects on Sparking Plugs. It has been shown that the design of a plug must be suited to the type of ignition system used for an engine. Thus, in transistorized and capacity discharge systems which give a relatively high voltage rise rate with a short duration spark it has been found that the rate of plug electrode wear is reduced and that there is a greater resistance to fouling. The arc duration must, however, not be too short or the engine performance may suffer. High output voltages are recommended since they provide a voltage reserve, which gives longer plug life.

The following table* gives some interesting statistics for the three different types of ignition system.

* See reference number 5 at end of this chapter.

Characteristics of Different Ignition Systems

Designation	Basic system type	Secondary capacitance* (mmF)	Voltage output at 1,000 rev/min (kV)	Average spark duration (μs)†	Average voltage rise time (μs)
A	Conventional automotive	65	27	1,600	70
B	Transistor switched	70	28	3,000	120
C	Capacitor discharge (using standard automotive coil)	65	30	300	50
D	Capacitor discharge (using low inductive coil)	50	29	12	6

* Includes average value of coil capacitance, leads, and spark plugs.
† At 1,200 rev/min. road load.

Electrode Gap Plugs. These plugs have a small gap in their central conductors, so that the spark has to jump between the gap before it can reach the plug electrodes. The purpose of the gap is to isolate the high tension during its build-up so that there is no current flow across the insulator deposits—which act as a shunt resistance R across the electrode gap P in Fig. 254. It is claimed that this type of plug, which is specially suited to ignition needs of two-cycle engines has a longer working life than the continuous electrode one. However, it has now been shown that the capacity discharge system plug is superior to the electrode gap type and that in two-cycle engines using the petroil lubrication system the surface discharge plug, with a suitable ignition system, as mentioned on page 182, gives freedom from plug fouling.

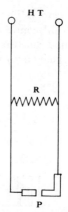

Fig. 254. Illustrating the shunt resistance property of plug insulator deposits

The Battery. There have been some interesting developments in lead–acid batteries in recent years which have resulted in improved performance, weight reduction per watt-hour, longer battery life and easier maintenance. Instead of the earlier exposed inter-cell metal connectors, these are mostly concealed, leaving only the two end terminals exposed. One American make of car battery has its end terminals in recesses in the side or the casing, the terminals being threaded to take the screwed cable connectors. It is claimed that this method eliminates terminal post corrosion.

An appreciable reduction in battery weight has been achieved by making the battery case of a specially tough plastic material of the co-polymer group; the cell separators and cover are also of the same material. As this co-polymer cannot be united by solvents or adhesives a thermal weld method is employed to unite the cover and case. Perhaps the most important improvement is that in which the cell connectors pass through the cell partitions instead of over their tops, a method which much reduces the electrical resistance of the connectors. Thus, in a typical case the resistance of the 'through the cell' connector was only one-sixth that of the conventional method and the reduction in voltage loss increased the battery voltage by 0·7 volts.

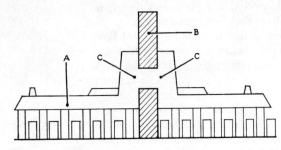

FIG. 255. Showing the improved method of car battery construction. (A) battery plate connector. (B) cell partition. (C) sides of plate connector

Fig. 255 shows one of the battery connectors A through the cell partition B, after its two sides C had been welded together. As compared with the conventional battery the new cell connector battery gave 25 per cent more cranking turns to the engine, at 15 per cent higher speed and for 15 per cent more cranking time. In the case of another make of battery of the same size, by using better materials for the casing, cover and separators and less non-active lead in its construction, it was possible to reduce the weight by 25 per cent, thus improving performance. Another innovation, by Delco–Remy is the vacuum sealing of all battery cells to stop additional water vapour, oxygen or other reactive gases, from entering the cells under the cover or its casing.

Improved Battery Topping-up Methods. Modern batteries dispense with cover plugs over each cell and instead usually employ a single trough running almost the length of the cover, in which distilled water is poured and by suitable devices fills each of the cells to their correct level. Examples of such methods include the Exide Auto-fil, Lucas Autolok and Dagenite Easifil. The principle of the latter method is shown in Fig. 256. The top of the battery case is at T and there is a single cover for all of the cell openings below. When the cover is off (A) glass balls seal the openings in the vent holes of the cells. During the topping up operation (B) an air lock is formed when the acid rises to the bottom of the filling hole in each cell. When the cover of the trough is replaced the acid rises to the correct level (C) and the openings are sealed off.

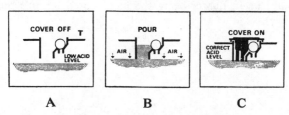

FIG. 256. The 'Easifil' method of topping-up battery cells in one operation

Headlamps. The four-headlamp system, originating in the U.S.A., has since been adopted in some British cars, but more recently in the square cover glass, instead of the previous circular, form. The square type gives a larger reflector area, with improved road illumination. Hitherto, the conventional headlamp with a pre-focus bulb and separate reflector was used in four-headlamp systems, but since then, Lucas have produced a rectangular sealed beam type headlamp, which gives a greater light-collecting area, due to the absence of the bulb holder. In this new type the main and dip filaments are accurately located in relation to the parabolic reflector, to give better road illumination.

A low-wattage lamp with its holder is located in an adapter at the rear of the sealed headlamp unit to provide parking and sidelamp lighting. The light from each of these lamps is visible through a window panel in the reflector.

The Quartz Iodine Lamp. Known also as the Quartz Halogen type, this later headlamp bulb has been developed over a period of years to give all of the functions of conventional tungsten-type bulbs in headlamps, but in addition much better road illumination. It was initially used on competition cars for high speed night driving on Continental roads. Basically this lamp consists of a bulb with a tungsten filament, but filled with an inert gas containing a small amount of iodine vapour. When current is applied to the filament it reaches an appreciably higher temperature than that of previous bulbs, namely about 600° C with a correspondingly brighter filament. The higher operating temperature makes it necessary to use a stronger heat-resistant transparent material for the bulb, itself, quartz being the most suited for this purpose (Fig. 257).

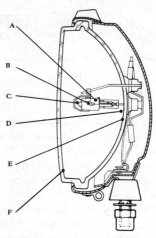

FIG. 257. The Lucas sealed beam quartz-halogen spotlamp. (A) halogen envelope. (B) correct focus filament. (C) shielded filament. (D) centrally-masked reflector. (E) sealed beam unit. (F) optically designed lens for wide flat-topped beam

When in operation the filament emits tungsten atoms which pass into the surrounding gas and there combine with the iodine vapour to form tungsten–iodide. The convectional movement of the gas in the bulb causes the tungsten-iodide to pass to the hotter regions in the bulb where a certain reaction occurs which results in the tungsten and iodine separating, the former element returning to the filament and the latter to the gas in the bulb. This process of reaction and reverse reaction continues all the time the filament receives current. The quartz iodine bulb lamp has the following advantages over the previous type of tungsten filament lamp:

LATER DEVELOPMENTS IN ELECTRICAL EQUIPMENT 395

(1) Its bulb internal surface does not blacken as does that of the conventional lamp after much service. Neither does its filament burn away.
(2) It gives a more powerful light beam for the same wattage. Thus in the Lucas fog lamp of the conventional type, the filament gives about 80,000 candelas, whereas the quartz iodine fog lamp give about 110,000 candelas. For higher wattages the intensity of the beam is much greater within the temperature limits for the filament.
(3) The beam is much whiter than for the conventional lamp.
(4) Owing to its much smaller dimensions the quartz iodine bulb can give better optical beam properties. It is thus possible with the improved reflector and lens system available to produce a flat-topped beam with a minimum of light scatter.

With the four-headlamp system used on cars one of each pair of lamps can be dipped and the other lamp main beam switched off, but in the two-headlamp system with two filaments in each lamp it has proved difficult to employ the quartz iodine system until more recently, when a solenoid method has been used to move the filament to give the dipped beam. In another design of 7-in. diameter headlamp two reflectors and bulbs are used to give the main and dipped beam as required.

Fig. 258 shows a method of producing a dipped beam by using the rearward filament of a two filament quartz iodine bulb. It employs a hood or shield above the lower beam filament while leaving the upper beam for normal driving, when switched over from the dipped filament. This system can be designed to give a flat-topped beam.

Automatic Headlamp Control. With reference to the American headlamp control system described on page 289, more recently a new type of operation, due to Lucas, has been used which augments the car's dipped beam and adjusts, automatically, by means of shutters the beam intensity of an independent projection type of headlamp. In this system the projected light doubles the driver's viewing distance on the nearside but does not dazzle the driver of the oncoming vehicle.

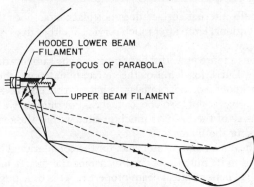

Fig. 258. Method for producing a dipped beam by using two filaments in the quartz iodine lamp bulb

Known as the Autosensa, it combines a light receiver and projection unit in a single lamp housing (Fig. 259). The receiver unit comprises a receiver lens, a photo-conductive cell and lens system, together with an electronic amplifier controlling a linear actuator which moves the receiver aperture shutter. The projector unit has a quartz–halogen bulb, ellipsoidal reflector, projector lens and a second shutter which controls the aperture of the projected light beam.

In operation the Autosensa is normally connected for use only with the lower dipped beams. When these are 'On' the lamp projects a rectangular pattern high intensity light beam. With the approach of a vehicle an image of its headlamps, produced in the Autosensa's receiver is projected on to the surface of the photo-cell (Fig. 260). An amplified signal from the photo-cell energizes the linear actuator which moves a shutter across the aperture until the shutter 'homes' on the image. A second 'shutter, attached to the first one, moves in unison with it and controls the offside edge of the projected beam which, otherwise, would dazzle the oncoming driver. At the same time the nearside part of the road is fully illuminated.

With a pair of 7-in. sealed headlamps the seeing distance with lowered beams in use to pass another vehicle approaching the driver's car is about 200 ft. When augmented by the Autosensa

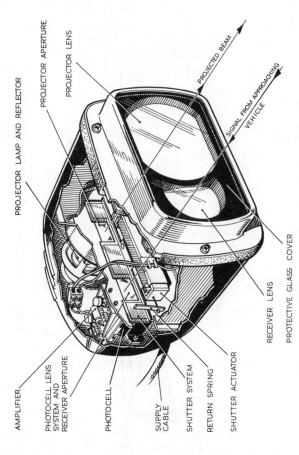

Fig. 259. Components of the Autosensa headlamp

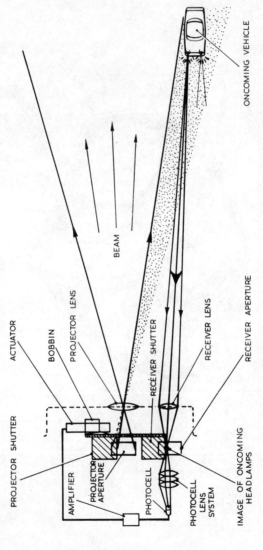

Fig. 260. Illustrating the operation of the Autosensa headlamp.

LATER DEVELOPMENTS IN ELECTRICAL EQUIPMENT 399

device this distance is about 400 ft. When closing on a vehicle ahead the tail lamp of same is usually bright enough to actuate the shutters at a distance of 200 ft. thus avoiding dazzle in the vehicle's rear mirror.

Automatic Switching of Sidelamps. The problem of switching on the front and rear sidelamps of a vehicle when the driver is away can be solved in several ways. Perhaps the most simple method is to use an electrically-operated clock, taking its current from the battery. The clock is fitted with adjustable 'On' and 'Off' switching devices for the sidelamp circuits. It is, however, necessary to make adjustments at intervals to take account of the varying times of sunrise and sunset.

Another method employs photo-cells which, when the daylight intensity falls to a certain value, operates a circuit having an amplifier to switch 'On' the sidelamps in the evening and 'Off' in the early morning.

Yet another, but more complicated system employs a circuit having 2 transistors, 8 resistors, 2 diodes, 2 capacitors and 2 relays to switch the sidelamps 'On' and 'Off' at set times. It also contains a manually-operated switch to cut out the automatic systems if not required.

Day and Night Signal Lamp Illumination. Experiments on car signal lamps, e.g. those for direction indicators and brake warning purposes have shown that the day illumination should be at least twice that needed for night driving conditions; for this reason modern cars now use a two-level system. In the Lucas Night–Day signalling system the day illumination intensity is obtained from a constant voltage bulb while that for night conditions is reduced by connecting resistors in the circuit. Fig. 261 shows the circuit diagram for the system. It employs a small relay unit with three sets of contacts, together with the voltage-reducing resistors that use a printed circuit. The relay operating coil is energized by the lighting switch control, so that when the side and tail lights are switched on resistors are inserted in the signalling circuit. Since, however, the reduced current drawn from the circuit under night conditions would alter the operation of the flasher

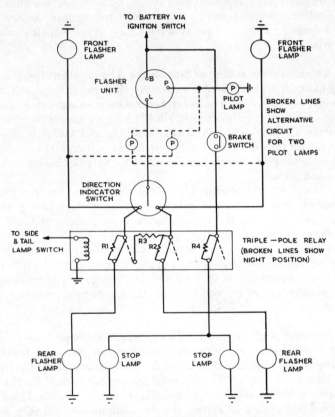

Fig. 261. Circuit diagram for the Night–Day signalling system for side and tail lights

unit a further resistor is connected in parallel with the lights to give an additional load on the flasher unit, thus correcting the frequency of the flasher.

Later Model Flasher System. The earlier Lucas type FL5 unit has since been replaced by the 8FL one, the circuit for which

LATER DEVELOPMENTS IN ELECTRICAL EQUIPMENT 401

is shown in Fig. 262. In operation, when the direction indicator switch is turned to the right or left the appropriate signal lamp bulb lights up immediately, the current flowing *via* the flasher terminal B, the normally closed contacts, the metal ribbon, the metal vane and terminal L. The current through the metal ribbon causes the latter to heat and expand; this allows the vane to relax and thus

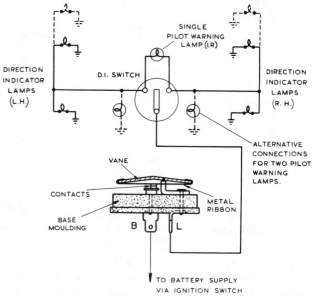

FIG. 262. Circuit diagram for direction indicator flasher system

open the contacts. The signal lamp bulbs are extinguished and the ribbon then cools and applies tension to the vane, thus causing the contacts to be closed again and the cycle is then repeated. The snap action of the vane provides an audible indication of the flasher operation, while a pilot lamp on the instrument panel gives visible indication. If one signal lamp fails, audible warning ceases but the pilot lamp and the other lamps remain on but do not flash.

The flashing rate is from 60 to 120 per minute and the percentage of lamp 'On' time is from 50 to 70 per cycle.

A Transistorized Flasher Unit. The Simms direction flasher unit instead of using vibrating contacts to interrupt the signal lamp current employs the electronic switching method, thus avoiding the use of any moving members. A power switching type of transistor is energized by a clock pulse from a multi-vibrator circuit to provide the current interruptions. The components are mounted on a printed circuit mounted on a metal plate to provide a suitable heat sink.

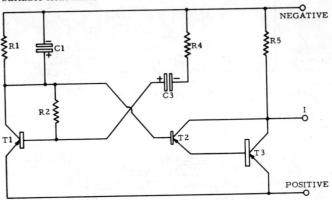

FIG. 263. Circuit diagram for the Simms' transistorized direction indicator flasher system

Fig. 263 shows the circuit diagram for the 12-volt flasher system, the principle of the operation of which is based upon an astable multi-vibrator circuit, in which two transistors $T1$ and $T2$ generate the clock pulse which is fed to the power switching transistor $T3$. The lamp current passes through $T3$ which operates in cascade with $T2$. The ratio of the light-to-dark period is determined by the resistor $R4$ and capacitor $C3$. Adjustment of the flashing frequency is made during manufacture by variation of the resistor $R2$ which is specially selected for each unit. The capacitor $C1$ prevents any incoming impulse from altering the flashing rate. A red warning light on the instrument panel indicates the operation of the system.

Transistorized Variable Speed Windscreen Wiper. This later design of windscreen wiper provides for the variation of the wiper blade speed operation. It uses a transistorized delay unit which can be used with the single-speed self-switching type of wiper unit. This enables the wiper stroke cycle to be varied from the normal frequency, by means of a knob control with a graduated dial, from 1 to 10 seconds. The Lucas delay unit consists of a double-contact change-over relay, a transistor governing this relay and a variable resistance-capacitance circuit which determines the interval between the cycles. A rigid-type printed circuit board is used. During the period that is set by the control knob switch the capacitor becomes charged to a voltage level at which the transistor switches on and allows the capacitor to discharge and energize the relay. The battery supply is thus connected to the wiper motor *via* the relay and the motor starts to operate. The circuitry is so arranged that the capacitor now discharges, switching off the transistor and cutting out the relay. At one end of the wiping cycle the wiper stops under the normal action of the limit switch. At this point the capacitor again begins to charge and the operation is repeated.

Plastic Fibre Optics. A method of illuminating objects from a single distant light source by means of a special type of 'optical' cable has been brought to the production stage for cars, in the U.S.A. The 'cable' consists of a number of strands of a plastic material, namely, polymethyl-methacrylate, in a sheath of a transparent polymer having a lower refractive index. These strands are enclosed in an opaque outer flexible jacket. If a light source is placed at one end of the cable the light rays travel in a zigzag path by internal reflections along the cable and thus provide a source of illumination at the other end of the cable. Further, the cable can be bent or twisted, without affecting its optical conveyance properties. Fig. 264 shows an assembly of nine optical fibres within a cable and illustrates its illumination of nine different positions. In effect the cable assembly acts as a light conductor and the number of optical fibres can be increased to give greater light intensities at the exits. The Croften light guide cables are supplied with 16, 32, 48 and 64 individual fibres.

This method obviates the use of separate small electric bulbs and provides a lighted area wherever desired, e.g. on instrument panels, around panel switches, behind instrument faces, and in inaccessible

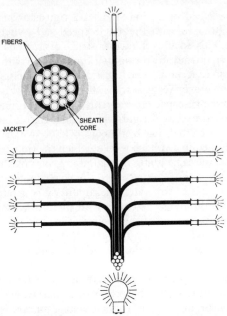

Fig. 264. Plastic optical fibre distant illumination system. The upper left diagram shows a section through a 19-fibre optical flexible cable

places for ordinary small light bulbs. Only one main light source is needed for the light cable and this is usually a lamp bulb with a reflector on the opposite side to the inlet of the cable end. Mechanical fittings are available to fasten the cable, but it is pointed out that these must not be compressed too much in tightening, as otherwise there will be light losses at the exit ends.

Impulse Type Tachometers. A more recent development in automobile engine revolution indicators is based upon the electrical impulses which occur when the contact breaker of the ignition system breaks the low tension circuit. In the case of the Smith tachometer these impulses are counted by means of a transistorized printed circuit contained within the instrument, which in turn influences a D.C. voltmeter to which a pointer is attached. The scale of the pointer's dial is graduated in r.p.m., from zero to 8,000 r.p.m. The accuracy of this tachometer is unaffected by variations in engine timing, sparking plug gaps or contact breaker gap settings.

The Diesel engine version of this tachometer depends for its timed pulses upon a small camshaft generator, the lobed wheel of the rotating member of which is located close to a magnetic pick-up unit, which is electrically connected to the tachometer on the instrument panel.

Impulse Type Speedometers. Instead of the conventional flexible shaft driven speedometer, the shaft drive can be dispensed with and in its place a small pulse generator mounted on the gearbox to take its drive from the same gearing which operated the flexible shaft. The speedometer then becomes an electronic type speed indicator, operating on the same principle as the Diesel engine tachometer. Not only does this type simplify the generator to instrument connections, but it is particularly suited to large vehicles where the gearbox is located at a relatively long distance from the instrument panel. In the case of vehicles with two-speed rear axle drives, the Smith Electromag speedometer has a compensator in the instrument head, such that the action of the driver in changing from one axle ratio to the other also operates a solid state switch which connects a correction circuit in the speedometer.

PUBLICATION REFERENCES

1. *Principles of Transistor Circuits*. S. W. Amos (Iliffe Books Ltd.).
2. 'The Use of Alternators on British Cars'. L. E. Edwards and A. W. Winley, *Proc. Inst. Mech. Engrs.*, 1967–68.
3. *Transistor Regulators*. Training Chart Manual. Delco Remy (AC–Delco Ltd., Dunstable, Bedfordshire).
4. *Transistor Ignition Systems Handbook*. B. Ward (W. Foulsham Ltd.).
5. 'Characteristics of New Ignition Systems'. R. C. Teasel and R. D. Miller, *Proc. Inst. Mech. Engrs.*, 1967–68.
6. 'Ignition and Spark Plug Requirements'. R. C. Teasel and R. D. Miller, *Proc. Inst. Mech. Engrs.*, 1967–68.

INDEX

A. C. Delco and Delco Remy
alternators, 366, 370
cold-starting ignition system, 378
distributor unit, 376
sparking plugs, 212, 219, 220
transistorized ignition system, 385
Alcomax, 207, 208
Alkaline batteries, 254 *et seq.*
Alkaline cell, 254
Alloys, magnet, 207, 208
Alni, 207, 208
Alnico, 206, 208
Alternators, 115 *et seq.*, 366 *et seq.*
   advantages, 115
   applications, 116
   C.A.V., 124
   Chrysler, 122, 123
   cooling, 117, 121, 122, 124, 373
   Delco-Remy, 116, 120, 367, 370
   diodes, 117, 369
   drive, 117
   Lucas, 116, 117, 368 *et seq.*
   outputs, 115, 118, 120, 121, 122, 123, 124, 367, 371
   rectifiers, 117
   regulators, 118, 120, 121, 122, 368, 374
   transistors, 118, 120, 369 *et seq.*
   types, commercial, 116, 120, 122, 124, 370 *et seq.*
   stator, 117, 373
   slip ring, 117, 121, 373
Alternator components, 121, 371, 372
   developments in, 366 *et seq.*
Aluminium oxide insulator, 214
American Autronic-eye headlamp control, 289
   beam aimer, 277, 278
American circuit-breakers, 358, 359
   convertible car hood, electrically-operated, 339
dual headlamps, 278, 279
flashing light indicators, 181, 182
fuses, 357
hardtop, electrically-operated, 334
headlamps, 265, 267, 277
instrument panel, 293, 294
instruments, 293, 296, 297, 299, 301, 304, 305, 314
ignition units, 165, 166, 168, 170, 174, 176, 184
seats, front, electrically operated, 336, 337
window, electrically-operated, 334, 335
Ammeter, 22, 307, 308
Annular spark gap, 143, 144
Atomic structures, 360
Auto-Lite Corporation, 243
Auto-Lite regulator adjustment, 109
Auto-transformer, coil, 153
Automobile electrical systems, 14
   lamp bulbs. *See* Bulbs, lamp
Automatic headlamp beam deflector, 288
Automatic ignition timing controls, 174 *et seq.*
Automatic rear mirror control, 344
Automatic voltage control. *See* Regulators
Auto-pilot, 342
Autosensa headlamp, 396
Autronic-eye, 289

Ballast resistance, ignition 154, 155
Base, transistor, 185, 361
Battery, 17, 20, 67, 233 *et seq.*
   acid. *See* electrolyte
   density measurement, 244, 245
   alkaline, 233
   armoured plate, 238
   capacity, 237, 240, 241, 242, 245
   charging methods, 249 *et seq.*, 258

INDEX

Battery—*continued*
  charging system, 17, 20, 67 *et seq.*
  chemical reactions, 233, 234, 235, 255
  cycling method, 247
  construction, 235, 236, 238, 391, 392
  density. *See* Specific gravity
  discharge rates, 240, 241
  electrolyte, 234, 235, 243, 257
  efficiency, 242
  faults, 252
  high rate charging, 250
  internal resistance, 248, 249
  lead-acid, 233 *et seq.*
  negative plate, 233
  performance graphs, 239, 256
  plate materials, 233
  positive plate, 233, 234
  principle of, 233, 234, 235
  ratings, 240
  separators, 234, 236, 237
  short time rating, 240
  specific gravity, of acid, 235, 243, 244
  sulphation of plates, 251, 252
  temperature effect on, 242
  testers, 244, 245, 246
  later developments in, 391
  topping-up methods, 392
  vent plug, 236, 237, 238
  voltage, of cell, 235
Beam setter, 276
Bendix Folo-Thru pinion drive, 51, 52
Bendix pinion, inboard, 50
  outboard, 40
Bendix pinions, 48, 50
Bimetallic strip voltage compensator, 92
Bimetal arm fuel gauge, 299
Bimetal arm temperature gauge, 304
Bimetal circuit breakers, 357, 358
Boost charging, batteries, 250
Bosch compensated voltage regulator, 106
  current and voltage regulator, 107, 108
  dynamo, 79
  regulator adjustment, 108, 109

  regulator units, 106
  semi-conductor (variode) regulator, 109, 110
  sliding-armature motor, 60, 61
  sparking plugs, 216, 225
Bourdon pressure gauge, 306
Brake warning light, 283
Breakaway torque, 32, 59
Breaker. *See* Contact breaker
British Standard sparking plugs, 224
  cables, 349
Brushes, 44, 76, 81
Brush loading, 44, 81
  materials, 81
  spring pressures, 76, 81
  sparking, 81
Bucking coil, 102
Buick, 267, 289, 290, 291, 295, 310, 341
Buick headlamps, 267, 290
  automatic control, 289, 290
  speedometer, 310
Bulbs, lamp data, 268, 274, 278, 285, 286

Cables, 315, 346 *et seq.*
  battery, 348
  colour identification, 315
  connections for, 355, 356
  current ratings, 347, 348
  high tension, 348, 349, 350
  sizes, 347, 348, 350
  types of, 347, 348, 349
Cable wiring harness, 354, 355
Cadillac, 174, 177, 289
Cadmium test, battery, 246, 247
Cam-angle (contact breaker), 167, 168
Cam dwell, 167
Capacitance, of ignition windings, 136
Capacitance, in suppressors, 230
Capacitance spark component, 136
Capacitors, 376, 381, 386
Capacity, battery. *See* Battery capacity
Carbon deposits, sparking plugs, 211, 219
Carburation, effect on starting, 35
Carburettor motor starter device, 63

INDEX

Car interior heating systems, 333
  ventilating systems, 333
C.A.V. alternating current generator, 124
C.A.V. battery, armoured type, 238
C.A.V. nickel-cadmium battery, 258
C.A.V. starting motor, 62
Centrifugal ignition advance mechanism, 174, 175
Ceramic insulators, 214
  finish of, 216
Champion sparking plug, 225
Characteristic voltage curve, 87, 88, 90, 91, 100
Charging rate. *See* Dynamo charging rate
Chrysler, 52, 122, 297, 338, 342, 344
Circuits, electrical, 17, 19, 21, 23, 25, 26, 287, 288, 291, 314, 343, 351, 352, 353
Circuit breakers, 168, 357, 358
Cirscale, 313
Clevite Corporation, 193
Clocks, electric, 313, 314
Clutches, electromagnetic, 340
Coil and magneto system comparison, 147, 148
Coil ignition system. *See* Ignition, battery and coil system
Coil ignition, advantages and disadvantages, 150
  complete circuit of, 152
  components, 156 *et seq.*, 162
  considerations, 155
  current (primary), 148, 149
Coil ignition limitations, 180
Coil ignition, limiting speeds, 155, 180
Cold sparking plug, 219, 220
Cold-starting ignition system, 378
Co-axial starting motor, 61, 62
Collector, transistor, 185, 361
Columax, 208
Commutator, 44, 78
  brushes, 44, 78
Compensated voltage control (Co.V.C.), 87 *et seq.*
Constant current battery charging, 250
  potential battery charging, 250

Constant voltage control (C.V.C.), 83 *et seq.*
  unit for auxiliary circuit, 300
Condenser, ignition, 134, 152, 153, 166
  capacity, 134, 167
  construction, 166
  discharge magneto, 210
Contact breaker, 20, 134, 135, 156, 161 *et seq.* 376
  construction, 162, 164, 165
Contact breaker design, 164, 165, 376
  gap, 163
  speed limitation, 155, 163, 164
Continental (U.S.), 315
Control resistance. *See* Ballast resistance
Corrosion, battery terminals, 252
  sparking plugs, 211
Crankshaw and Arnold, 192
Current and voltage control regulator (C. and V.C.) principles, 98, 99
  designs, 98, 104, 106, 107, 109
Cut-in voltage. *See* Closing voltage
Cut-out, adjustment methods, 96, 108, 109
Cut-out, circuit arrangement, 69
Cut-out, closing voltages, 70, 96, 97, 103
  conventional diagram, 70
  contacts, 71, 113, 114
Cut-out contacts gap, 96
  drop-off voltage, 113
  principle, 21, 67, 68, 69
  temperature compensation, 92
Cycling, battery, life tests, 247, 248
Cycling test, battery, 247
Cyclometer, 310

Delco-Remy. *See* A.C. Delco
  C. and V.C. regulators, 104
  contact breaker unit, 165, 170
  distributor, 170
  H.T. coil, 160
  ignition advance mechanism, 176
  over-running clutch drive, 54, 55, 57
  regulator adjustments, 108
  vacuum control, ignition, 178
De Soto (U.S.), 335

Detonation, 172
Diesel engines, 19
 cranking horse-power, 42
 heater or glow plugs, 19
 starting of, 19, 36, 55
Diodes, 119, 185, 361, 362
Dip-and switch headlamp, 272
Direction indicators, 285
 flashing type, 285 *et seq.*
 semaphore, 285
Disruptive discharge, 132
Distributor, 20, 152, 156, 168 *et seq.*
Distributor cap, 162, 170
Distributor components, 162
 drive, 161, 162, 170
 erosion, 171
 rotor arm, 169
 system layout, 168
 tracking fault, 171
 ventilation, 171
Dodge (U.S.), 327
Door locks, electrical, 341
Driver control device, 341
Driver speed warning device, 341
Drop-off voltage, 113
Dry charge type battery, 253
Dynamos, 71 *et seq.*
 alternating current types. *See* Alternating current generators
 bearings, 77, 78, 79, 118, 121, 122
 belt tension, 81, 82
 charging rate, 73
 cooling of, 93
 commutator, 78
 construction, 76, 77, 78
 current outputs (rated), 73, 76 79, 80, 104
 data, 80
 drives, 81
 electrical specifications, 113
 lubrication, 78, 121
 potential curves, 72
 regulators. *See* Regulators, dynamo
 shunt, 75 *et seq.*
 shunt characteristics, 82, 83
 speeds, 67, 70, 73, 74, 76, 80, 113, 114
 temperature effects on, 91, 93
 third brush type, 69, 70 *et seq.*
 two brush type, 71, 75 *et seq.*
 types, 76, 77, 79
 ventilation, 77, 79, 117, 121, 122, 124

Earth, negative pole method, 29, 360
 positive pole method, 15, 29, 360
 return system, 15, 29
Electric Autolite Company, 184
Electrical applications, summary, 27
Electrical cooling fan, 331
Electrical energy analysis chart, 28
 fuel pumps, 328, 329, 330
Electrical gauges, engine, 294 *et seq.*
Electrical speedometer, 311, 312
Electrical door locks, 341
 speed control device, 341, 342
Electrical system, complete, 24, 26
 schematic layout, 25
Electrical systems, to understand, 16
Electrically-operated convertible car hoods, 339
 door locks, 341
 door windows, 334
 front seats, 336
 hardtop, 339
 reserve petrol supply, 344
 transmissions, 340
Electrode-gap sparking plugs, 391
Electrons, 131, 360
Electronically-controlled headlamp, 289, 290
Electronically-controlled rear mirror, 344
Electronic spark explanation, 131
Emitter, 185, 361
Endurance test, ignition equipment, 145
Engine cranking power, 243
Engine cranking torque, 31, 32, 242
Engine electrical gauges, 292 *et seq.*
 instruments, 309 *et seq.*
Engine starting system, 17, 36 *et seq.*

Fan, radiator cooling, 331, 332
Ferranti battery tester, 246
Flasher system, transistorized, 401, 402

INDEX

Flywheel magneto, 209, 210
Ford ignition unit, 179
Ford (U.S.) 41, 65, 66, 234, 288, 304, 337, 338, 339, 381
  contact breaker unit, 179
  starting motor, 65, 66
  wiring diagram, 26
Forward bias, 362, 363
Frequency coil. *See* Bucking coil
Fuel gauges, 295 *et seq.*
  bimetal type, 298, 299
  coil indicator type, 295, 296
  constant current type, 291
  thermal strip type, 298
Fuel grade, and ignition timing, 179, 180
Fuel level warning light, 300
Fuses, 356, 357
  American practice, 357

Gap, contact-breaker, 163
  cut-out, 96
  regulator, 96, 97
  sparking plug, 223
Gas turbine, fuel igniters, 183, 226
Gauges, cooling water, 301
  electrical circuits for, 314
  fuel level, 295
  pressure, 305
General Motors Corporation, 289, 341
Germanium, 110, 360, 364
Glass seals, sparking plugs, 217
Glossary, English-American, 11
Glow plugs, 19
Graves headlamp, 270, 271
Ground. *See* Earth
Guide-matic headlamp, 289, 290, 291

Headlamps, 262 *et seq.*
  automatic control of, 289, 395
  adjustments of, 264, 266, 268, 274 *et seq.*, 289, 290, 393
  aiming lugs, 267, 268
  alignment, 274, 280
  alignment of beam, apparatus, 275
  bulb wattages, 268, 274, 278, 285
  dip-and-switch type, 272
  dual type, 278, 279
    alignment, 279, 280

electronic control, 288, 289
focus of, 262
focusing, 263
fog, 281
Graves, bulb type, 270, 271
light intensity graphs, 273
light patterns on wall, 274, 275
long beam type, 282
non-dazzle beams, 270 *et seq.*
optical principles, 262, 263, 269, 282
pass-lamp type, 282
pre-focused bulbs for, 264
quartz-iodine, 393
recent developments in, 393 *et seq.*
reflectors, 262, 264, 268
sealed-beam type, 265, 268, 269
twin-filament bulbs, 272, 273
twin-unit type, 278
Heat sink, 117, 122, 364
Heater, or glow plugs, 19
High speed engine ignition systems, 155, 156, 163, 164, 180, 188 *et seq.*
High tension. *See* H.T.
High voltage spark, 130, 132
  theory, 131, 132
  production, 132
Horn, electrical, 316, 324
  double-type, 325
  high frequency, 324
  low frequency, 324
  relay types, 327
  wind-tone, 326
Hot sparking plugs, 219, 220
Hot spot, 218
H.T. cable specifications, 348, 394
H.T. coil, 19, 20, 132, 156, 157 *et seq.*, 379 *et seq.*
  components, 157, 158
  construction, 158
H.T. coil current consumption, 20, 159
H.T. coil, oil-filled, 159
H.T. voltage, charge pressure and temperature effects, 141
  factors influencing, 141
  throttle opening effect, 141
H.T. voltages, 130, 137, 138, 140, 143, 144, 145
  voltage, rate of increase, 137, 138

Humber, 319
Hydraulic analogy to electrical system, 14
Hydrometer, 244
Hynico, 208

Ignition advance angle, 127, 128, 172
Ignition agent, 128
  effect on indicator diagram, 126, 127, 128
  endurance test, 145
  Ford flywheel generator system, 130
  interference, radio and T.V., 228
  make-break spark system, 129
  requirements, 146
  ring fuel injection system, 194
  Ruhmkorff indication coil system, 129, 133
Ignition requirements, starting motor, 34
Ignition system, 17, 18, 125 et seq., 374
  capacitor-discharge type, 386
  characteristics of, 390
  coil, 132, 135, 152 et seq. 374 et seq.
  Delco-Remy, 190
  electronic, 183 et seq., 379
  for high speed engines, 155, 163, 180, 379, 380
  fully-transistorized, 381
  later developments in, 379 et seq.
  low-voltage, 181
  Lucas, 189, 190, 378, 886, 394, 395, 403
  magneto, 135, 146, 195 et seq.
  piezo-electric, 190, 191
  timing, 171 et seq. (see also Timing of ignition)
  transistor-assisted, 381
  transistorized, 183 et seq.
  warning light, 284
Ignition, utility test, 145
Impulse ratio of spark, 132
Indicator diagram and ignition effect, 125, 126, 128
Indium, 360
Inductances in suppressors, 230
Induction coil, 133
Inductance spark component, 137, 138

Inertia pinion, starting motor, 51, 52
Instrument panels, printed circuits, 364
Instruments, electrical, 292 et seq.
  ammeter, 307
  clock, 313
  constant voltage supply, 300, 301
  design trends, 294
  fuel gauges, 295 et seq.
  lights, 283, 292, 294
  low level fuel warning device, 300
  oil pressure gauges, 305, 306, 307
  oil pressure lights, 305, 306
  panels, 293, 294, 365
  printed circuits, 294, 364
  temperature gauges, 301
  speedometers, 309, 405
Insulated wiring system, 15
Interference, radio and T.V., causes, 226, 227, 228
Interference, suppression, 230, 231
Ionization process, 131
Ionized sparking gap, 143
Ions, 131, 360

Joules, 137, 139 et seq.

Lamp signal illumination control, 399
Lead, tetraethyl, 221
Leaded fuel, effect on sparking plug, 121, 222
Lead salts, 221, 222
Leyland, 312
Life tests of batteries, 247, 248
Light intensity graphs, 273
Lighting system, 22, 261 et seq.
Lighting system, brake warning lights, 283
  direction indicator lamps, 285, 286
  energy demands, 261, 262
  fog lamps, 281
  headlamps, 262 et seq.
  instrument and indicator lights, 283
  lamp wattages, 274
  pass-lamps, 282

INDEX

Lighting system—*continued*
 rear red reflectors, 283
 reversing lamps, 284
 side lamps, 282
Lincoln (U.S.), 41, 293, 315, 335, 336
Load control, ignition advance, 176, 177
Locked torque, 32
 test, 45, 46
 values, 43, 45, 59, 65
Lodge sparking plugs, 225, 227
Low voltage ignition systems, 181
Lubricating oil, starting effects of, 33
Lucas, 51, 52, 64, 76, 94, 96, 98, 116, 163, 169, 188, 189, 202
 alternating current generator, 116, 117, 366 *et seq.*
 battery, 238
 camshaft magneto, 201, 202
 compensated voltage, regulator, 94
 contact-breaker unit, 165, 378
 current and voltage regulator, 97, 98, 102
 cut-out, 96
 direction indicator, 286
 distributor, 169
 dynamos, 76, 77
 electrical fuel pump, 329, 330
 electronic ignition system, 188, 189, 190, 380 *et seq.*
 headlamps, 264, 268, 393 *et seq.*
 headlamp aligning equipment, 276, 277
 motor-driven radiator fan, 331, 332
 oil-filled H.T. coil, 159
 reserve petrol supply device, 344
 starting motors, 64, 65
  motor pinion drive, 51, 52, 53
 windscreen washer, automatic, 322
  wiper, two-speed, 320
 wiring diagram, simplified, 351, 352

Magnet steels and alloys, 207, 208
Magneto, 135, 146, 195 *et seq.*
 advance and retard method, 199
 Bosch, 195, 196
 camshaft, 201, 202
 condenser discharge type, 210
 current change graph, 198, 206
 flux change graph, 198, 206
  changes, 197, 205
 flywheel-type, 209, 210
 ignition timing, 199, 200, 204
 revolving magnet type, 200, 201, 209
 rotating armature type, 195, 196
 polar inductor type, 200, 204, 205
 speed range, 199
 voltage change graphs, 198, 206
Magneto magnet sizes, 208
Mercury (U.S.), 41, 168, 234, 357
Mica insulator, 214
Micro-circuits, 364, 368
Mirror-matic mirror, 344
Miscellaneous electrical equipment system, 24
Motor, seat-moving, 336, 337, 338
Motor starting. *See* Starting motor
 window lifter, 334 *et seq.*
 windscreen wiper—*see* Windscreen wiper motor
Moving iron ammeter, 308
Moving iron indicator, 303

Nash (U.S.), 280
Negative bias, 361
Negative earth system, 29, 30, 360
Nickel alloy electrodes, 213
Nickel-cadmium battery, 254, 258
Nickel-iron alkaline battery, 254 *et seq.*
No-load, starting motor, 37, 45, 46, 65, 66

Oil deposits, sparking plugs, 211, 219, 220
Oil-filled H.T. coil, 159, 160
Oil pressure indications, 305
 gauge, 305, 306
 warning light, 305
 test, 306
Oldsmobile (U.S.), 296
Opus ignition system, 386
Overcharging of battery, 252
Over-running clutch, solenoid-operated, 55
Over-running clutch, starting motor, 48, 54, 57, 61

INDEX

Patterson and Campbell, 140
Piezo-electric crystals, 191
Piezo-electric ignition system, 190, 191, 192
Plastic fibre optics, 403
Platinum electrodes, sparking plug, 213
Plug, heater, *see* Heater plug
Plug, low voltage, 182
Plug, sparking. *See* Sparking plug
Plug, surface discharge, 181
Polar inductor magneto, 200, 204, 205
Positive earth system, 15, 29, 30, 360
Power-operated car hoods, 339
Power-operated seats, 336 *et seq.*
  windows, 334
Pre-ignition, 218, 219
Primary circuit, 135, 152, 153, 154
Primary coil, 131, 132, 135, 153, 154, 157, 158, 159
Printed circuits, flexible, 365
  rigid, 364
Pulse generator, magnetic, 384
Pump, fuel, 328, 329
  centrifugal, 330
  diaphragm, 328
PZT-ceramic element, 193

Quartz-iodine headlamps, 393 *et seq.*
Quick-action cable connector, 355

Radiator cooling fan, electrical, 331
Radio receivers, 226, 230, 345
  ignition interference with, 228
  general interference with, 226, 227, 228, 230
Rambler (U.S.), 40, 45
Rectifiers, silicon diode, 117, 121, 122, 124, 361, 362
Reflectors, headlamp. *See* Headlamp reflectors
  rear, red, 283
Refrigerated air car ventilation systems, 334
Regulation graphs, dynamo, 74, 83, 88, 90, 91, 100, 112

Regulator, compensated voltage (Co.V.C.), 87 *et seq.*
  constant voltage (C.V.C.), 83 *et seq.*
  contact metals, 101
  current and voltage (C. and V.C.), 68
  double contact, 86, 87, 104
  dynamo, 68 *et seq.*
  gaps, 96, 97
  ideal type, 68
  single contact, 84, 87, 104
  single core cut-out and voltage regulator, 103, 107
  temperature compensated, 75
  temperature-voltage data, 93, 114
  third-brush method, 68
  transistorized, 367 *et seq.*
  typical designs, 94, 95, 98, 104, 106, 107, 109
  Tyrill, 68
Renault, 82
Resisting torque, 32
Resistivity, sparking plug fouled, 221
Resistors, 361
Reverse bias. *See* Negative bias
Reverse current relay. *See* Cut-out
Reversing lamps, 284, 285
Ring ignition process, 194
Rotating armature magneto, 195, 196
  magnet magneto, 200, 201, 209
Rotor arm, distributor, 169
Rubber coupling type inertia pinion, 32, 33
Ruhmkorff coil, 129, 133

Screenjet, 322
Sealed beam headlamp, 265, 266, 267, 268
Secondary circuit, 135, 152
  coil, 131, 152, 157, 159
Semi-automatic starting scheme, 64
  transmissions, 340
Semi-conductor, 109, 110, 302, 359
Semi-conductor thermometer unit, 302
Side lamps, 282, 399
  automatic switching, 399

INDEX

Silicon-diode rectifiers, 115, 117, 184, 185, 361, 364
Sintox, 183
Simms flasher system, 402
Smiths, 307, 311, 340
Smith's automatic transmission, 340
  battery, 236
  electric clock, 314
Smitzs, W. B., 181
Snap-on connectors, 355, 356
Society of Automative Engineers (S.A.E.), 218, 247, 349, 356
  sparking plug specifications, 218
  H.T. cables specifications, 349
  snap-on terminal, 356
Solenoid hold-in winding, 57, 58
  pull-in winding, 57, 58
Solenoid-operated starting pinion, 48, 54 *et seq.*
Spark, capacitance component, 136
  component considerations, 138
  electronic theory, 131, 132
  energy data, 139, 140
  energy formulae, 130, 137
  gaps, 138, 139
  gap voltages, 144, 145, 146
  high voltage. *See* High voltage spark
  inductance component, 137
  intensity, 141, 142
  mixture strength effect on, 140, 141
  oscillatory nature of, 137, 138
  production, 132
  voltage. *See* High voltage spark and speed effect, 148, 149
Sparking plug, 138, 142, 152, 156, 211 *et seq.*, 387 *et seq.*
  American standard, 218
  cold type, 219, 220
  construction, 212, 216
  developments in, 387
  electrode materials, 213
  electrode temperature, 142
  erosion, 211, 213, 221, 222
  fouling, 220, 221
  gap location, 388
  gaps, 138, 238, 388
  glass sealing method, 217
  hot type, 218, 219, 220

insulator materials, 213, 214 *et seq.*
insulation resistance, 142
  leaded fuel, effects on, 221, 222
  life of, 225
  materials, 212, 213
  multi-point, 217
  operating conditions, 211
  platinum type, 213, 225
  polarity, 225, 366
  sealing, 216, 217
  shell, 213
  standard specifications, 218, 224
  tapered seat, 218
  temperatures, 211, 218
  temperature ranges, 218
  thermocouple type, 225
Spaulding, H. G., 184
Speeds, ignition, limiting, 155, 156, 163, 164, 180
Speeds, dynamo. *See* Dynamo speeds
  engine, ignition systems, 148, 149, 155
  ignition type, 313
  magnetic induction type, 309, 310
Speedometers, 309 *et seq.*, 405
  impulse type, 405
Speedometer, developments, 313, 405
  electrical types, 311
  horizontal scale type, 310, 311
Spotlight, 263
Starting Diesel engines, 19, 36, 55
Starting motor, 17, 36 *et seq.*
  axial-sliding type, 48, 59, 60
  breakaway torque, 32
  brushes, 44
  cables, 347
  circuits, 46
  co-axial type, 61, 62
  commutator, 44
  components, 64, 65
  compound-wound, 36, 37, 40
  cranking power, 42
  data, 43, 44
  drives, 48 *et seq.*
  driving torque, 32
  efficiency, 44
  formulae, 37, 38, 44

INDEX

Starting motor—*continued*
  locked torque, 32, 45, 46
  horse-power to drive, 42
  ignition requirements, 34
  lubricating oil effect, 33
  no-load test, 45
  performance graphs, 39, 42
  resisting torque, 32
  series-wound, 36, 37 *et seq.*
  shunt-wound, 36, 37
  specifications, 45
  speeds, 31, 34, 39, 42, 43, 45, 50, 66
  switches. *See* Switches, starting motor
  temperature effects, 33
  torques, 31, 32, 242, 243
  torque terms, 32
  three-pole type, 41
  typical designs, 64, 65
Static charge on tyres, 232
Steels, for magnets, 207, 208
Sulphation, battery, 251, 252
Suppressors, general, 226, 230
Suppressors, sparking plugs, 226, 227
Surface-discharge ignition plugs, 181, 182, 225, 226
Switches, ignition, 19, 152, 153
  magneto, 196, 204
Switches, solenoid, 47, 55, 56, 58, 59, 60, 62
  starting motor, 15, 17, 47, 55 *et seq.*, 60
  thermostatic fan, 331

Tachometer, engine, 313, 404
Tachometer, impulse type, 404
Tach-dwell meter, 168
Temperature compensator for voltage regulators, 92, 101, 120
  controlled radiator fan, 331
Temperature effect on battery, 242, 243
  effect on dynamo output, 91
Temperature, effect on engine starting, 33
  effect on voltage regulators, 91, 113, 120
Temperature, and electrolyte density, 243

Temperature gauges, 310
  of sparking plug electrode, 142, 211, 218
  range, sparking plugs, 218
Temperature measuring sparking plugs, 225
Temperature—voltage data for regulators, 93
Tetraethyl lead fuels, 221
Thermal runaway, 363
Third brush dynamo. *See* Dynamo, third brush
Three-point spark gap, 132, 143, 144
Ticonal, 207, 208
Timing, of ignition, 126, 127, 128, 171 *et seq.*
Timing, ignition, and indicator diagram, 126, 128, 172
Timing, ignition, advanced, 127, 128, 172, 173
  centrifugal advance mechanism, 174, 175, *et seq.*
  effect of engine speed, 173
  incorrect, effects of, 172, 174
  micrometer control, 179, 180
  retarded, 172
  vacuum control, 176, 178, 179
Torque, engine starting. *See* Starting motor torque
  locked. *See* Starting motor locked torque
Torque terms, 32
Torque-sensitive radiator fan clutch, 332
Transistorized generator regulators, 118, 119, 120, 367 *et seq.*
Transistorized ignition systems, 183 *et seq.* 379 *et seq.*
  system, without contact breaker, 188, 383
Transistors, 183 *et seq.*, 361 *et seq.*
Transformer, H.T. *See* H.T. coil
Transmission systems, electrical, 340
Trembler coil, 129, 132, 133
Tungsten contacts, 101, 164, 171
Tungsten electrodes, 213
Turbulence, charge, effect on plug voltage, 142
Two brush dynamo. *See* Dynamo, two brush

Utility test, ignition equipment, 145

Vacuum control, ignition advance, 176, 178, 179
Variode characteristic curve, 111, 112
Variode dynamo regulator, 109, 110
Vauxhall, 293, 303, 311
Ventilated dynamos, 77, 79
Venner silver-zinc battery, 260
Voltage, piezo-electric crystals, 192, 193
Voltage, charge turbulence effect, 142, 143
Voltage, cut-out, closing. *See* Cut-out closing voltages
Voltage, H.T. *See* H.T. voltages
Voltage regulators, method of adjustments, 109, *et seq.*
Voltage regulators, 68, 84, *et seq.*
Voltage regulator, single contact, 84, 87, 89
 double contact, 86, 87
 sparking, measurement, 143
Voltage surges, in regulators, 101
 in spark gaps, 137, 138

Warning lamp in alternator circuits, 369
Warning lights, brake, 283, 292
 direction indicator, 283, 285, 286, 292
 door locks, 341
 high headlamp beam, 283, 292, 293
 ignition, 284, 292, 293
 low fuel level, 300
 oil pressure, 283, 292, 293
 water temperature, 305
Watson, E. A., 94, 146, 163
Wattages, lamp bulb. *See* Headlamp bulb wattages
Windscreen wipers, 316 *et seq.* 402
Windscreen wiper, blades, 317, 318, 403
 cable rack type, 318
 central motor type, 320
 gearbox, 318, 319
 motor, 317, 318, 319
 transistorized variable speed, 402, 403
 triple blade, 322
 two-speed, 320, 321, 322
 variable speed, 320
 washers, 322, 333
 wiring diagram, 321
Wiring cables. *See* Cables
Wiring diagram, Lucas simplified type, 351, 352, 353
Wiring diagrams. *See* Circuits, electrical
Wiring harness, 354
 cable connectors, 355
Wiring of automobiles, 346 *et seq.*

Young, A. P. and Warren, H., 148

Zener, 363, 374, 381, 386